Mathematics in Nature, Space and Time

John Blackwood

# Mathematics in Nature, Space and Time

Waldorf Education Resources

Floris Books

First published as *Mathematics Around Us*,
and *Mathematics in Space and Time*, in 2006

This edition published in 2011 by Floris Books
© 2006 John Blackwood

British Library CIP Data available


ISBN 978-086315-818-6
Printed in China

# Contents

# Introduction

In a recent article in the *Sydney Morning Herald* (20 Dec 2001) John Metcalfe of the Parent-controlled Christian Schools is quoted as saying 'Children are taught that maths is a language for describing the world — a language that God created ...'

This is something that I too have felt for many years now, and feel that it is an approach that can go far, very far, if taken with some seriousness. It assumes there is a secret to be revealed in the book of nature and that the world is far more than a long-term probabolistic accident, or calculable through embarrassingly huge extrapolations which no practical engineer would dream of making. One can, of course, permit oneself any number of views, after all there is no reason that *materialistic* science should have a monopoly on objectivity.

## *Nominalism or the language of the gods?*

One view is that the world of mathematics is a *nominalistic* and abstract collection of convenient ideas with little, if any, reality of their own and having mere convenience and pragmatic value in relation to understanding the world 'out there.' Although commented on by some thoughtful folk, the fact that mathematics is so *extraordinarily* good at this usually escapes our attention.

Another view is that mathematics is — in more senses than one — a language of the gods. It could be contended that what our minds apprehend in the concepts of mathematics and geometry is the last pale residue of the active forming forces that 'make the world happen.' The view held here does not assume that the thoughts we have are intellectual conveniences, mere shadows of the mind, but rather a veritable doorway towards a growing comprehension of the workshop of nature. One can take the view that, not only are there gods, but it is also possible to take an interest in *how* they work. This has been my attitude

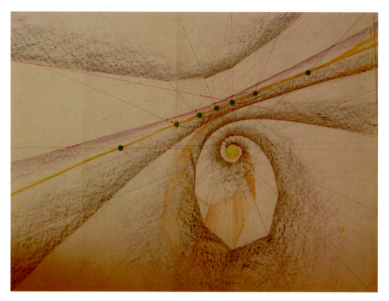

*Fig I.1  Generalized spiroid*

and I find myself affirmed in this through the sheer wonder of nature and the beauty of the subject we are dealing with. To me Shakespeare's 'pale cast of thought,' only applies to our current thin intellectuality, not to what the thought life might eventually achieve — as Rudolf Steiner has pointed out in his *Philosophy of Spiritual Activity*, 'with thinking we have grasped one small corner of the spiritual.'

No doubt there is a whole range of variations among these positions as well, and the whole thing could be debated endlessly. The amazement at the concordance of the mathematical concept with the carefully observed phenomenon can enable us to be very busy, intrigued and interested despite epistemological niceties — important though they no doubt are.

This book contains material from the main lessons I have given to Years 7 and 8. Each main lesson is done over approximately three weeks for one and a half hours each morning in our school, Glenaeon Rudolf Steiner School, in Australia.

Every teacher does these sessions differently and what is done ends up being unique to the group and to the teacher, to the place and to the time. However there seems to me to be a 'golden thread' to which we are all striving.

The Waldorf curriculum provides a constant challenge to every teacher. This challenge is to develop our work year by year for the sake of the students — and to develop ourselves through it too. I often wonder how we can expect our students to develop, work and grow if we ourselves are not striving to do the same. How can the kids grow if we are not? There has to be an equation here!

These notes are offered as a contribution to the mathematical themes.

I also thank the many students and friends whose work I have used in illustrations. Where they are known and accessible I have acknowledged this. My apologies if I cannot give personally all credit where credit is due.

It should hardly be necessary to suggest that these notes are only one individuals choice of material and that many would include quite other selections.

However, it is what I have worked with over some twenty-one years and it has been, over those years, of interest to some students and some colleagues — judging by the photocopies I have had to do!

John Blackwood

# 1. Mathematics in Nature

At this beginning of puberty there is an increasing demand from the young folk to be able to connect the thoughts they can have *about* the world *with* the world they actually experience. Mathematics, and geometry in particular, can now be seen in much that surrounds us in the miracle that is nature. Something dawning inside us meets the phenomenon outside.

If there is a correspondence in the educative life of the children with the general growth and development of humanities consciousness then the historical period reflected is early and pre-Renaissance. This need to personally connect these two sides of life emerged for the individual at that time, and emerges strongly at puberty in the students' life too. The battle for autonomous thought was fraught with difficulty with the dominant faiths of the time deeming themselves under threat by individual efforts of the likes of Copernicus, Galileo and Kepler. The awakening mind of the young person also challenges *us* at this time!

We re-member at this time in the students' life the kind of work, approach and exploration just beginning to explode onto the world in Central Europe, and world-shaking and world-shaping it was — the Renaissance. But we need to do it out of where we are now.

What follows is a pictorial outline of some of the themes and topics that I did in a particular main lesson at Glenaeon Rudolf Steiner School many years ago in attempting to cover the kind of work I believed belonged to this age. Much else could be incorporated of course, and there is no pretence of completion, but this was some of the content offered at that time.

The pages which follow take typical exercises in the sequence that they were offered at the time. Sometimes there are suggestions as to accompanying activity and sometimes the initial sequence of steps of the work are indicated.

*Fig 1.1 The title page of the students' main lesson book should hint at the subsequent themes and contents for the three week main lesson*

## Review of some skills

We reviewed a couple of simple geometric constructions: bisecting an angle and drawing a line at right angles to another given line — just to get started.

   This calls for compass and ruler. It always seems to me that careful use of the compass cannot be emphasized enough. A reasonable compass that does not wobble or skid, or widen spontaneously, as well as a sharp pencil, and a ruler with a clean edge (and not pockmarked by being banged against the edge of a desk) are essential. The ruler should be 30 cm long or more.

*Fig 1.2  Double rainbow over Sydney*

*Fig 1.3  Drawing a line perpendicular to another, bisecting an angle
— and a rainbow (compass work)*

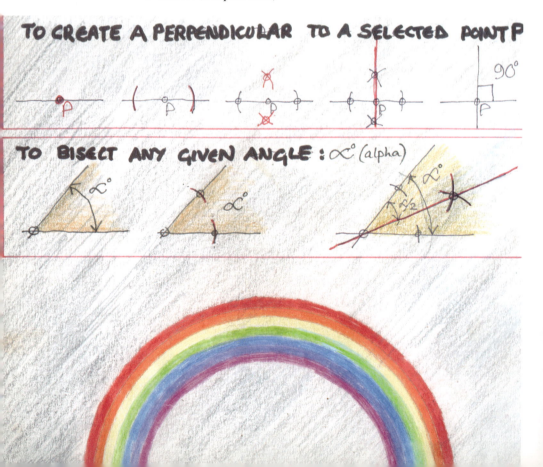

TO CREATE A PERPENDICULAR TO A SELECTED POINT P

90°

TO BISECT ANY GIVEN ANGLE : ∝° (alpha)

*Exercise 1   To draw a perpendicular to a given line*

1. Draw a line, *p*, and place on it a point, *P*, through which the perpendicular is to be struck.

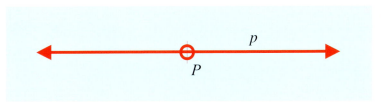

*Fig 1.4*

2. With a compass set at a radius of (say) 5 cm, and the pointy end set on *P*, mark the points, *A* and *B*, on either side of *P* as shown.

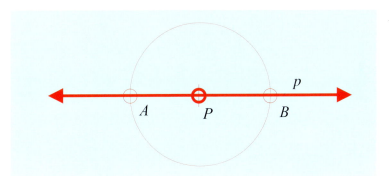

*Fig 1.5*

3. Now draw two arcs from *A* and *B* respectively, of a radius greater than 5 cm (say 7 cm) so that they intersect in *C* and *D*.

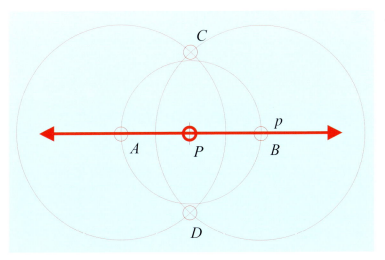

*Fig 1.6*

4. Join the points, *C* and *D*, and we have the line perpendicular to line *p* through the point *P* as required.

*Fig 1.7*

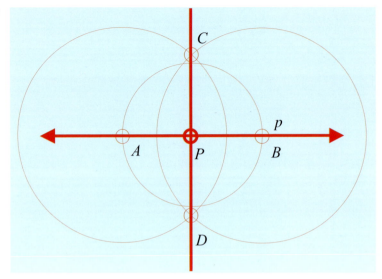

This is a further simple, but essential introductory exercise:

*Exercise 2   To bisect any given angle,* α

1. Draw two lines, *b* and *c*, intersecting in *A*, with α, the angle to be bisected (i.e. cut in half), between the two.

*Fig 1.8*

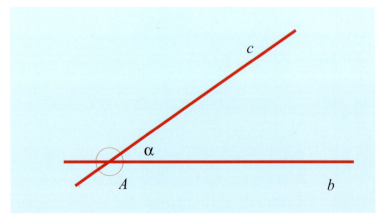

2. Select a radius, *AC,* of about 5 cm and set a compass to this radius. Put the compass point on point *A* and draw an arc creating points *B* and *C* on lines *c* and *b* respectively.

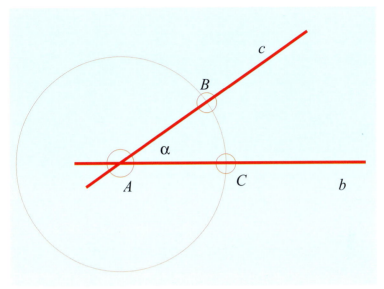

*Fig 1.9*

3. With compass at approximately the same radius draw arcs from *B* and *C* respectively to intersect at point *D*.

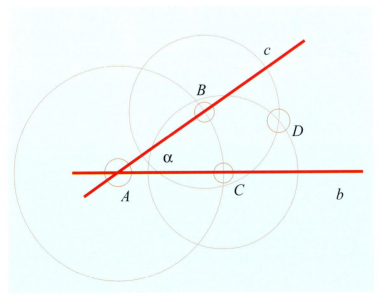

*Fig 1.10*

4. Finally, draw the line AD. This will be the required bisecting line,
   where ∠BAD is equal to ∠CAD
   i.e. ∠BAD = ∠CAD = β     hence 2β = α as required.

*Fig 1.11*

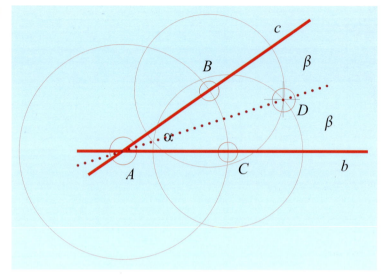

These above two constructions will be used in future diagrams
and will only be referred to briefly if needed.

### *Exercise 3  Rainbows*

But what is out there in nature that has a patently geometric
structure? Very many things if we look hard enough; and
a delightful candidate often appears against dark shower
clouds.

    This third exercise approximates the rainbow. It is not easy
to take photographs of these wonders as they span so much of
the sky. But this is an excellent exercise in drawing concentric
circles and then bringing the appropriate colours to them. Noted
that red is on the *outside* of the inner bright (primary) rainbow
and on the *inside* of the outer fainter (secondary) bow.

    Conventionally there are the seven colours. Richard Of York
Gained Battles In Vain is one mnemonic for the red, orange,
yellow etc.

Fig 1.12  A rainbow as semi-circle

1. Set up a horizontal straight line.

2. Mark a point about its centre.

3. Set a compass at a radius about half the line length.

4. Scribe *eight* semi-circles (which means seven spaces) from the centre, where each radius is (say) 3 mm larger than the one before.

5. Colour between the circles!

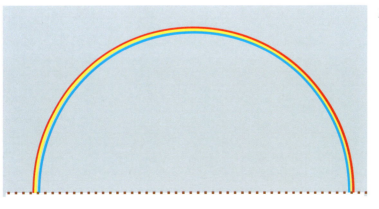

Fig 1.13

Do this on grey paper and finish with pastels — it can look startling. Above are only shown three such concentric circles attempting to partially emulate an actual rainbow as seen in Sydney early in the morning when the sun is low. To render this marvel to scale and with its magical luster is well nigh impossible.

## Circular forms

Where do we see circles? Drop a stone in a pond and circles of ripples emanate from the impact point as source.

### *Exercise 4   Circles, from points and lines*

We follow this rainbow drawing with an exercise that divides the circle into 16 divisions and which eventually appeared to give a series of concentric circles made from an array of near tangents.

1. First draw a line lightly across the paper through the middle *horizontally* and then, using the process described above for a right angle, construct a *vertical* line at 90° to it. Where they cross is point *O*.

*Fig 1.14*

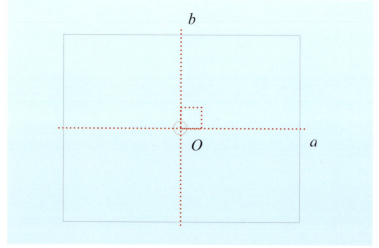

2. Bisect the right-angle on both sides. And bisect again all the eight new angles formed. This will create 16 divisions around the centre point, *O*. Now the angle between each line is

$$360 / 16 = 22.5°$$

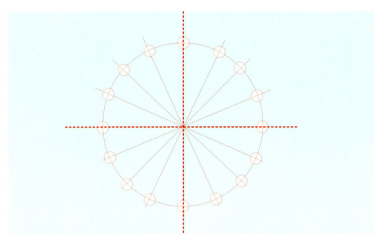

*Fig 1.15*

3. To obtain one of the concentric circles we now join (for example) every *fifth* point around the circle. It is easier to keep track if the points are numbered 1 through to 16.

4. Having done one such circle made from a family of (in this case) $2 \times 16 = 32$ lines, now try every sixth point and then every seventh and so on.

*Fig 1.16*

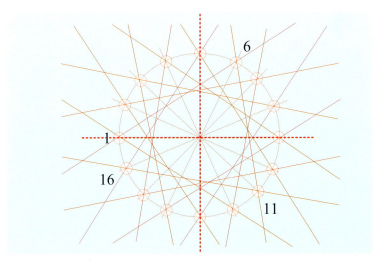

This will eventually lead to a family of approximate circles defined by series of tangents which are concentric to each other as is shown in Fig 1.17. It makes for an interesting picture if each resultant circle is formed of lines of a different colour. We thus obtain a *form* purely from a series of ordered lines. There will be more of this!

Note that the drawing indicates a suggestion of the lines not merely joining the points, but going *beyond* them. In essence a line is infinitely long, the two points merely defining its position:

Circles are everywhere, the whorl of the flower, the disc of the sun, the face of the moon, expanding ripples in the pond.

The next exercise explores this a little but it also enables us to see forms emerging that do not have the regularity of the circle but are still harmonious and symmetrical in themselves.

◄ *Fig 1.17 Concentric circles*

▼ *Fig 1.18 Circles looking down into a pool. Can you see them?*

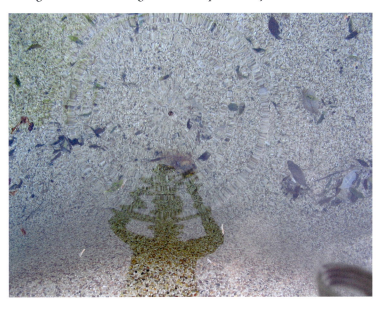

## Exercise 5  Asymmetric forms

The circle can be formed as in the exercise above (Exercise 4), but with a small modification of the construction, quite other forms can come about. Let us construct them first and then see if anything like them can to be seen around us.

1. Draw the perpendicular lines further *down* the page.

2. Now draw the equiangular radiants through point *O*. In this case 15° intervals have been chosen.

Fig 1.19

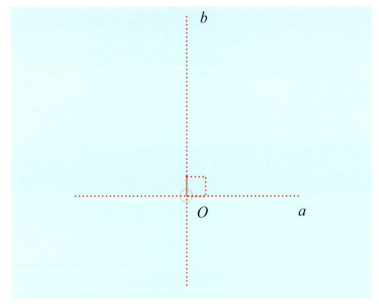

3. Then place a circle with its centre somewhat above the point *O*. Now mark and number (from 1 to 24) the points where the circle intersects the twelve radiants.

4. Join every fifth point with a line all the way across the drawing (see Fig 1.21). This will create the first curve.

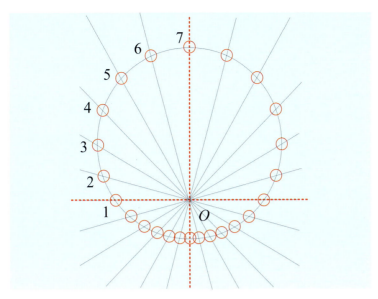

*Fig 1.20*

Then, as previously, join every second, third etc. point with a different colour: This gives a number of approximations to ovals formed from the tangent lines. A whole series of different *forms* appears.

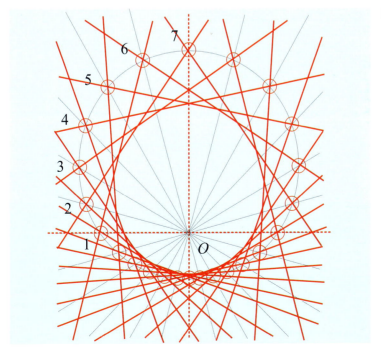

*Fig 1.21*

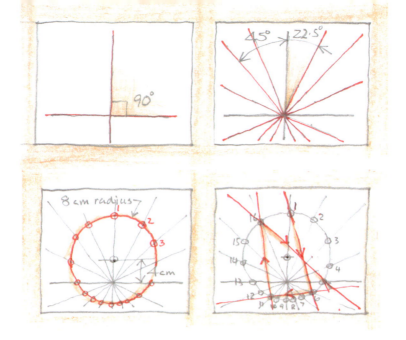

◄ *Fig 1.22 Asymmetric oval form*

▲ *Fig 1.23 Constructing the ovals*

5. The sketch in Fig 1.22 shows some of the whole family of these curves or ovals. Numerous constructions can follow as every 6, 7, 8, 9 etc. points can be joined. Within the centred circle these lines formed concentric circles (as we saw in Fig 1.17). But, when the circle is offset as in this case, a field of nested ovals emerges and it is interesting to challenge the students to see if they can discover any kind of order *outside* the circle. Teachers should try it first.

Note that with *even* numbered points you will have to have more than one starting point to get the full form. Odd numbers will eventually come back to the same starting point.

Do we see such ovals about us anywhere? The shapes look a bit like ellipses. Leaving aside whether they actually *are* ellipses for the moment we note that some egg forms look to all external appearances, very close to these.

The emu egg profile is a case in point. Accurate analysis shows it is close but not quite. Other eggs can be close too, and not just bird eggs. Local fauna in Australia, both platypus and echidna have nearly elliptical egg forms. This is a study in itself. Suffice to say that here is an obvious kinship between the phenomenon of an egg and conceptually devised oval (see Fig 1.24).

*Fig 1.24 Emu egg and an ellipse*

Still staying with the circle as a beginning it may be of value to see what happens when the circle is divided into six. Are there representatives of a sixfold character in nature? Little doubt about this one!

## Hexagonal forms

We drew a number of examples of sixfold symmetry. The students should be encouraged to find their own examples as much as possible. It can be amazing what a resource the class can be when many eyes (including parents and friends) are out there looking!

The basic hexagon is simple to construct. Bees are doing it all the time, so much so that artificial foundation for the cells in the

*Fig 1.25  Hexagons galore!*

TESSELLATES THE PLANE — GIANTS CAUSEWAY — THE HEXAGON IS FOUND IN QUARTZ CRYSTALS,

-IT IS SYMMETRICAL

-IT HAS SIX CORNERS OR VERTICES — QUARTZ CRYSTALS

-IT HAS SIX LINES

-IS MADE UP OF SIX EQUI-LATERAL TRIANGLES THE SAME SIZE

-CAN BE DRAWN IN A CIRCLE AND OUTSIDE A CIRCLE WHEN REGULAR.

HONEYCOMB

DIANELLA FLOWERS

FOUNDATION

SNOW — THE LARGER FULLER CRYSTALS FORM HEXAGONAL PLATES ARE ARRANGED AROUND THE NUCLEUS FROM SIX RADII

TOURMALINE

ISOPHYSIS IN THE GIANTS CAUSEWAY IN THE BEES HONEYCOMB IN THE PATTERNS THAT MAN LIVING CELLS TAKE, IN MANY FLOWERS OF THE LILY FAMILY

TASMANICA

AND IN THE INFINITE VARIETY OF SNOW FLAKES THAT FALL

hive can be made from stampings in beeswax and the bees will use it. Having had bees myself, it is fascinating to watch how the bees will build their cells both with and without the artificial foundation. Quartz crystals often show a good six-fold symmetry. Many flowers, especially the lily family, exhibit this closely as does the sun orchid flower of New South Wales.

*Fig 1.26 –28  Hexagons in crystal, flower petal and beeswax layout*

What do we do with compass, pencil and ruler to construct a regular hexagon? Easy.

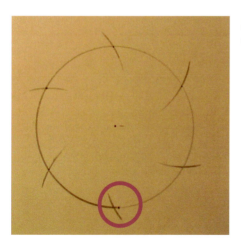

*Fig 1.29 Checking our accuracy!*

A good check on one's own accuracy is to start at any point on a circle, with the compass point. It is an accurate drawer that makes it so that the last arc does not overshoot or undershoot the original point! Try it.

## *Exercise 6   Constructing a regular hexagon*

1. Draw a circle, centre *O*, with a radius of about 5 cm. Draw a horizontal line through *O*, and mark the intersections of line and circle *AB*.

2. With compass set on point *A*, and retaining the circle radius, mark two arcs on the circle. Do the same from the point *B* as well.

3. Join *A* to *C* to *D* to *B* to *E F* to *A* and we have the hexagon.

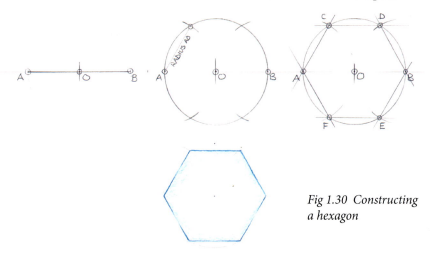

Fig 1.30  *Constructing a hexagon*

## *Exercise 7   Making a snowflake*

Another hexagonal form, which emphasizes the sixfold symmetry in nature is a snowflake.

1. To make a paper 'snowflake' take a thin A4 (letter) sheet of plain white paper and with compass draw a circle about 8 cm radius.

2. Construct a regular hexagon as in Exercise 6.

Fig 1.31

3. Cut out the hexagon neatly.

4. Now fold along a *diameter.*

5. Then fold to other corners until an equilateral triangle is formed.

6. From the point, or corner, which is at the centre of the original hexagon, imagine a line to the lower outer edge bisecting the 60° angle at the centre and fold again. This should leave a right-angled triangle.

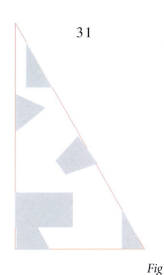

*Fig 1.32 Typical cut-out*

7. On the sides cut away the paper in any way you choose to mimic a snowflake (see Bentley and Humphreys amazing book *Snow Crystals,* for a large variety of snowflake photographs).

8. Unfold all the triangles, taking care not to tear the paper, and we have a great big 'snowflake'! The students usually love this one.

*Fig 1.33 Snowflake through cutaway paper*

## Spiral forms

Such exercises as the above let us see that in natural shapes there are mirrored what we can imagine in our minds geometrically, or mathematically. Does this mean the 'mathematics' is inherent in the expressions and widths of nature? Can it be otherwise? Are ideas mere nominal abstractions? Perhaps questions for later! It is certainly a question asked by many a scientist.

In the first instance the forms are simple and easily constructed. Now we come to forms which are *curved* in particular ways. Are such forms out there in nature all around us as well? Consider first a simple spiral.

## Archimedes spiral

The spiral of Archimedes could also be called the rope spiral (or the geometry of the raffia place mat). This is a spiral which is easily made with a piece of string. Start by holding one end of the string or rope, and wind a tight circle around until the string overlays the held end, and keep on and on winding.

This spiral has the characteristic that for each full turn the spiral gets larger by the *same* amount each time. It has a linear character. It increases stepwise forever — if the rope (or raffia) is long enough!

*Figs 1.34, 35  Rope stay for cannon — an Archimedean spiral*

*Exercise 8   Spiral of Archimedes*

This spiral can be drawn by the students as shown.

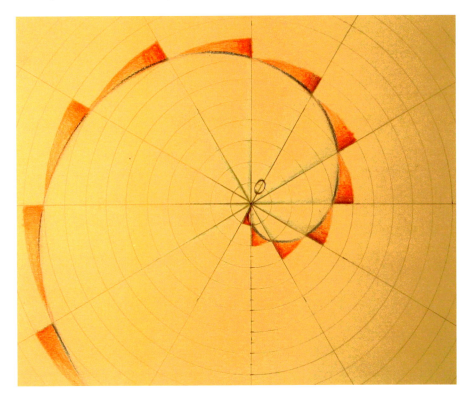

*Fig 1.36*

1. Draw vertical and horizontal axes.

2. With centre *O* draw concentric circles increasing in (say) 5 mm steps.

3. Draw radial lines every 30° (say) around *O*.

4. Now, starting at the centre, draw alternately steps out *radially,* and then around each arc for 30° step, then radially again and around and so on, alternately.

This was one particular and well-defined spiral. But it is not actually common in nature, as far as I can tell. If you find such a spiral, please share it with me.

## Exercise 9   Spirals galore

At this point it might be good to look around and see if the students can find the *general* spiral form in as many artifacts as possible. A very few samples are shown here. Make sketches.

The students should look to see how many examples they can find round about them, and then sketch them, from seashells to galaxies, water vortices to a head of hair (do men and women, boys and girls, differ in this one?).

*Fig 1.40  Vortex*

*Fig 1.37  Tunisian snail shell (Terry Funk)*

*Fig 1.38  Epitonum species sea shell*

*Fig 1.39  Ammonite from Frieberg*

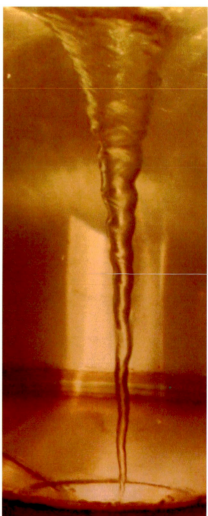

Within the image the following handwritten labels appear:

Spiral Sea Shells
TURBO
NAUTILUS
The Spiral of the WATER Vortex, the whirlpool and bath plug vortex.
Spiral in the hair crown of the head.
GALAXY

*Fig 1.41*

## The equiangular spiral

There are a number of other kinds of spiral. One is the *equi-angular* or *logarithmic* spiral. Interestingly a simple example of the equiangular spiral can be drawn by working with hexagons again as we see a little later. So we build with circles and hexagons. But we need a couple of preliminary construction methods.

### Exercise 10   Bisecting a line segment

We will need a method to bisect a given line segment. This can, of course, be done with a ruler by measurement but it is geo-metrically more elegant to use the construction shown here using compass and straight edge, and it is a useful skill to have.

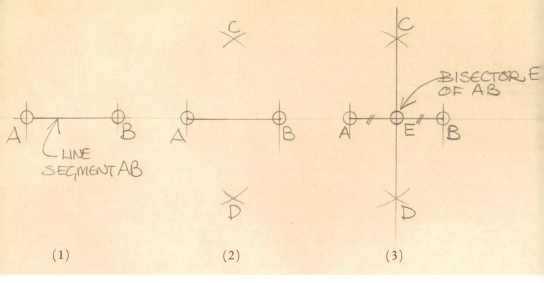

(1)                (2)                (3)

Fig 1.42 *Bisection of a line segment*

1. Draw the line segment, *AB*, which it is desired to bisect.

2. Set a compass at a radius approximately equal to the distance *AB*. Describe arcs at *A* and *B* so that they intersect at points *C* and *D*.

3. Now join points *C* and *D* and note where this line crosses *AB*. This is the point *E* which is placed so that *AE* is equal to *EB*, hence bisecting *AB* as required.

*Exercise 11   An equiangular spiral through a series of hexagons*

We now construct a series of hexagons, one within the other, getting smaller and smaller but in an ordered fashion. One could of course go the other way and draw larger and larger hexagons too, for the series goes on indefinitely, getting larger outwardly and getting smaller inwardly — never in fact actually reaching the centre.

1. Draw a circle of (say) 10 cm radius. On this circle construct a hexagon as shown in an earlier exercise (Exercise 6).

2. Bisect each of the six sides (Exercise 10).

3. Join these six bisecting points with six lines. This forms another, smaller hexagon. Bisect the sides of this new smaller hexagon.

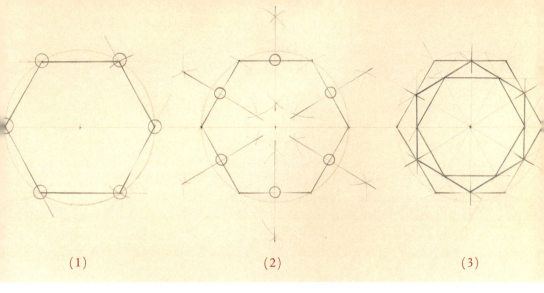

(1)                    (2)                    (3)

*Fig 1.43 Constructing a hexagon, bisecting the sides, constructing a further hexagon, and so on alternately. The hexagons get smaller and smaller.*

4. Join these midpoints and we have a third, yet smaller, hexagon.

5. Now pick out the *isosceles* triangles as they decrease in size going in an anticlockwise direction as shown in Fig 1.44.

6. Continue this process a number of times as the triangles approach the centre of the original circle. The sequence of triangles forms an easily constructed equiangular spiral, and many can be seen in the same diagram.

Various ways of further elaboration can give interesting effects. The students usually amaze us with their imagination.

*Fig 1.44 A family of equiangular spirals*

The more general (algebraic) expression for such a spiral is a little more complicated. Developing these expressions belongs to analytic geometry that we do in Year 11. It is interesting that the first set of coordinates we come to in *this* Year 7 main lesson are *polar coordinates* — although this is left unsaid as far as the students are concerned.

We notice that our spirals, the one derived from decreasing hexagons, and the more general expression, both attempt to converge upon the centre of the diagram.

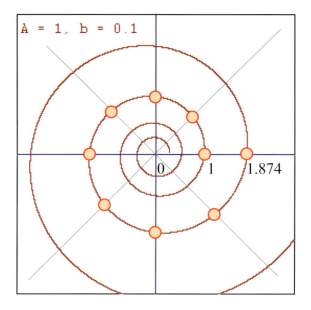

*Fig 1.45  An equiangular or logarithmic spiral using a True Basic program*

As the hexagons decrease it becomes harder and harder to draw them. The point to notice is that, mentally, we can *imagine* them (sort of) continuing forever, towards *a point at infinity,* on the page, but never seeming to be able to reach it. So we have what could be called a *local* infinitude.

For the general expression $r = Ae^{b\theta}$ it would be the same. If we programme a computer to plot this curve and simply let it run with continually *decreasing* values (i.e. negative) for $\theta$ it should never stop! (My program eventually comes up with 'overflow' if I don't limit it).

Likewise for *increasing* hexagons and angle *θ*, the spirals would expand outwards indefinitely. This says something quite important, and to hint at it to the younger students, namely that the geometric drawings we do are no more than fragments and actually imply that such drawings (in our minds at least) span the entire infinite. This actually applies to *all* geometric drawings, and in a certain way it means we can never draw a complete geometric drawing! Older students learn to *use* this idea. We can, after all, only try to *imagine* a line, for instance, at infinity. Now *that* is being holistic!

Do we see such spirals around us? Yes! Students need to look, find and share their discoveries in this area. A few examples are shown here including the famous nautilus, a shell used almost as an icon in this and other design work!

*Fig 1.46 Nautilus section*

Look along the axis of any cone shell and we see the spiral form. It is interesting to note that most cone sea-shells have a particular handedness or chirality.

*Exercise 12  Chirality or handed-ness check for sea shells*

1. Find examples of 'right-hand' spirals (that is, starting from the centre these turn clockwise). Are many sea shells dextrose, or right-handed? Check all the sea shells you can find. The answer may surprise you.

Fig 1.47  *Right-hand spiral shell. Start from the pointy end and ask: which way do I have to turn to go along the spiral of the shell away from me?*

2. Find examples of 'left-hand' spirals (that is, starting from the centre these turn anticlockwise).

Can you find any? Get the students to look — hard! For there are *very* few out there. The lightning whelk (*Busycon perversum*), a sea shell from the Gulf of Mexico, Yucatan, is one of the very few left-handers.

Fig 1.48  Busycon perversum — *left-hand spiral  (JB)*

There are few exceptions. Ask the students to look through any book on sea shells and see if they find *any* anticlockwise (or left-hand) spirals. They will find very few. Why this is the case is still a mystery.

Further examples are a beautiful pyrites fossil from Russia, an opalised spiral shell and an operculum shell, often seen on beaches around Sydney, Australia.

*Fig 1.49  Opalised shell (from Ann Jacobsen)*

*Fig 1.50  Operculum or trapdoor of a Turbo shell (from a local beach in Sydney)*

*Fig 1.51  Pyrites fossil (from David Bowden)*

In what follows we find that often we can see *interacting* spirals, both clockwise and anticlockwise for spirals appear in much more than only sea-shells.

First we look, though, at specific number sequences and slowly build a picture of spirals in the plant world.

As the hexagons get smaller, they decrease in the *same proportion*. That is, we multiply each successive diameter by the *same* number (less than 1). This is called the *common ratio* and defines a particular kind of sequence. However, there are other kinds of series which have particular importance in nature.

## Fibonacci numbers and sequences

One is the famous Fibonacci sequence. This was given by Fibonacci in about 1202 in his treatise *Liber abaci* concerning rabbits. And goes like this:

1,  1,  2,  3,  5,  8,  13,  21,  34,  55,  89 .......

*What is the next term after 89? Find the next five terms and state an expression for any term $F_n$ where n is the next term after the two terms n – 1 and n – 2.*

### Exercise 13   *Fibonacci numbers — the celery stick and others*

Where do we observe these Fibonacci numbers occurring in our world? In very many places actually. The humble celery stick, if a section is cut near the ground, hints at two different spirals, *one* in one direction and *two* in the other, *between* the flesh of the stems and thus demonstrating an instance of two consecutive Fibonacci numbers 1 and 2.

The spirals, interestingly, are what are *not* there materially, but *between* the flesh of the stalks. There is a big story here somewhere ....

Many plants have two sets of spirals — one clockwise the other anticlockwise, sometimes hard to see. The numbers of these spirals are often consecutive Fibonacci numbers.

One can make a 'print' of the celery section quite easily. Try it.

1. Obtain fresh celery plant.

2. Section horizontally the stick of celery with a sharp knife near the base.

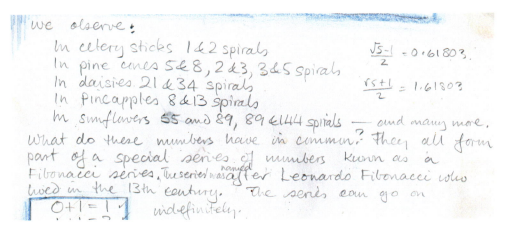

we observe:

In celery sticks 1 & 2 spirals
In pine cones 5 & 8, 2 & 3, 3 & 5 spirals
In daisies 21 & 34 spirals
In pineapples 8 & 13 spirals
In sunflowers 55 and 89, 89 & 144 spirals — and many more.

$$\frac{\sqrt{5}-1}{2} = 0.61803.$$

$$\frac{\sqrt{5}+1}{2} = 1.61803$$

What do these numbers have in common? They all form part of a special series of numbers known as a Fibonacci series. The series was named after Leonardo Fibonacci who lived in the 13th century. The series can go on indefinitely.

$$0+1=1$$

*Fig 1.52 Finding the Fibonacci series*

3. Hold the stems reasonably firmly and 'ink' or paint the cross section with solid colour. Carefully wipe off excess colour.

4. Now, on clean paper, make a 'print' of the section. It should look something like Fig 1.53.

*Fig 1.53 Celery cross section print*

Note that the colour shows up where the plant flesh is *not* but indicates its structure (or form). You may have to try this a few times! And then seek the double spiral. It is not easy to see ....

*Fig 1.54  Grass trees* — Xanthorea australis

Also the cross section of the *grass tree,* made just after fire, shows a strong suggestion of interacting spirals.

See if the students (and teacher!) can figure out how many there are and which way they are going. Again it is not easy.

*Fig 1.55  Grass tree section after a fire*

## Fibonacci spirals

The above are only two examples hinting at a hidden order. To mirror what nature appears to be doing we can find a way to construct such spirals geometrically. A very neat way is described quite fully by Peter Stevens in his book *Patterns in Nature*. The method we have used here for Fig 1.56 follows in the next major exercise.

### Exercise 14   Constructing a pair of Fibonacci spirals

Suppose we want to construct *five* equispaced spirals in a clockwise direction (yellow) and *three* in an anticlockwise direction (red) as in Fig 1.56.

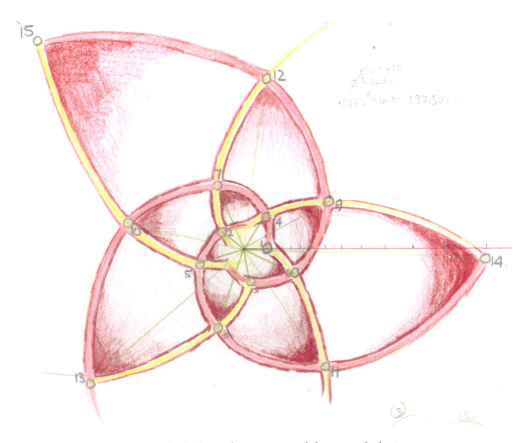

*Fig 1.56  Five spirals clockwise from centre and three anticlockwise*

1. Draw a horizontal line across the middle of the paper. On the centre of this line lightly mark a point *O*. Draw a circle 10 mm radius with *O* as centre.

2. We now draw a series of concentric circles about *O*. The radii about *O* are determined by what we call a 'multiplier' or common ratio, common because it is used again and again. In this case we could use 1.2 as our multiplier. Starting with a radius of 10 mm, to calculate the next term, we simply multiply by 1.2, (i.e. $10.00 \times 1.2 = 12$) and then 1.2 again and again. Hence the first few radii are:

   10
   12
   14.4
   17.28
   20.736
   24.8832
   29.85984
   35.831808
   42.99816960
   51.59780352
   61.91736422
   74.30083706
   89.16100448
   106.99322053
   128.3918464

3. Since we can hardly draw with pencil and ruler to much better than half a millimetre we could have rounded off as we went for our calculator usually retains the accuracy whether set to round off or not. We need to go to a radius of about 100 mm and round off to *one* decimal place as below.

   10.0
   12.0
   14.4
   17.3
   20.7

24.9
29.9
35.8
43.0
51.6
61.9
74.3
89.2
107.0
128.4

4. Draw concentric circles with all these radii, centre *O*, to cover the page. This is a good compass accuracy practice for this age group.

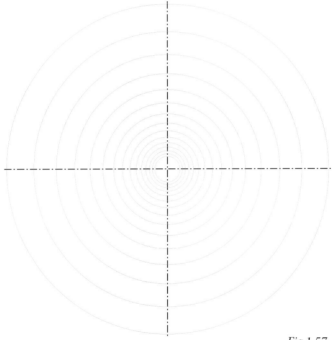

*Fig 1.57*

5. The next step will have to await proper justification until later. Suffice for the moment to say that, just as we can have a Golden section or ratio, so is it possible to have a Golden angle. The Golden section is explored a little later. Now we use the Golden angle itself. This angle is very close to 137.5° (the exact value can be calculated as shown in the panel — for those interested).

*Golden angle*

Assume the Golden section is

$$(\sqrt{5} + 1)/2 = 1.618033989...$$

as far as most school calculators will go. Now divide the full circle of 360° by this number i.e.,

$$360°/1.618033989 = 222.4922359°$$

and subtract this from 360°. This gives 137.577640° and this, to half a degree, is 137.5°. This is the angle we require, as the ratios of the angles about a point centre are:

$$360° \; : \; 222.4922359° \; : \; 137.577640°$$
$$1.618033989 \; : \; 1 \; : \; 0.618033989$$
$$(\sqrt{5} + 1)/2 \; : \; 1 \; : \; (\sqrt{5} - 1)/2$$

or, in words, the ratio of the whole to the larger part is the same as the ratio of the larger to the smaller part.

We could use a normal protractor now, starting from the horizontal line and marking sequentially anticlockwise lines from the centre. But it is far easier to make a single paper angle 'protractor' of 137.5° and step around the centre with this angle, drawing rays to the periphery of the drawing with each anticlockwise turn of this Golden angle.

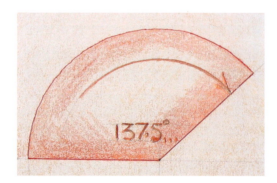

*Fig 1.58 Golden angle 'protractor'*

6. Starting from the smallest 10 mm radius circle, mark a point, on the horizontal line right of centre, where the radius cuts the circle. Mark this as point 1. Then point 2 will be the intersection of the circle with radius 12 mm and the line at $1 \times 137.5°$ anticlockwise from the centre. Point 3 will be the intersection of the circle with radius 14.4 mm and the line at $2 \times 137.5°$ anticlockwise from the centre. And so on, round and round in an anticlockwise direction ... continue until you have run out of points (there were 15 calculated) or have covered the page!

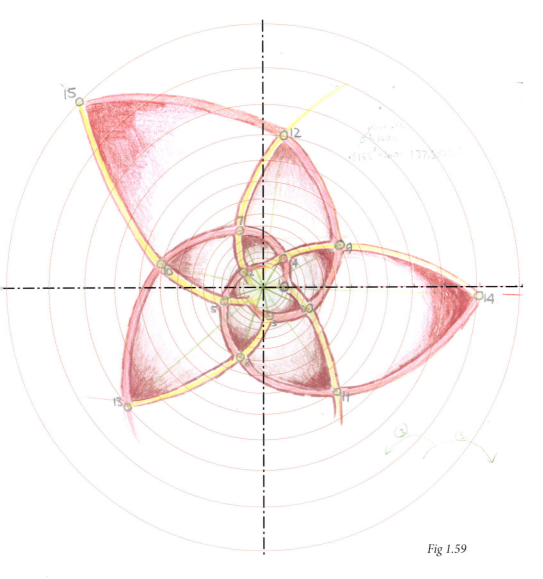

*Fig 1.59*

7. To obtain the spirals themselves we join a series of the numbered points. One series is used for the three anticlockwise spirals and a different series for the five clockwise spirals. For the five clockwise spirals join points (5, 10 and 15) and (2, 7 and 12), (4, 9 and 14) and (1, 6 and 11) and finally (3, 8 and 13). Note these are sequences of every *five* points.

   For the three anticlockwise spirals we join three sequences of every *three* points. Mark both series of spirals in lightly.

8. Finally firm in, in free hand, the curves themselves as below. Do it for yourself first before leading Class 7 through this exercise.

*Fig 1.60*

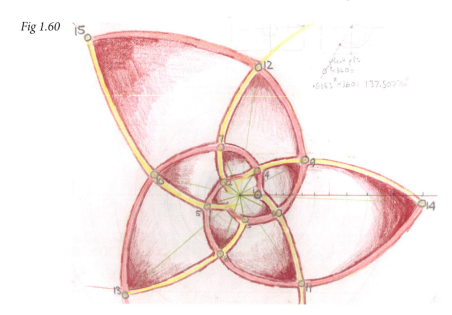

Note, that as for the celery section, we can fill in the areas *between* the spirals. In some plant forms it is suggested that it is what is *filled in* that we see physically. It is the mind's eye that discerns the *spiral* architecture as such, for we are actually describing something that is not materially manifest. For me it is enough however, in this case, that the students see and draw the beauty of the forms.

   We have seen how this Fibonacci series of numbers is glimpsed in the way related to spirals in plant form. We can even draw such spirals. Another appearance is in the human body

*Fig 1.61 Golden ratio dividers (model by Roger McHugh)*

where certain proportions approximate to a particular number that is intimately related to this peculiar series.

I built a kind of proportional measuring device based on a special pair of dividers which a friend, Roger McHugh, had made many years ago. His elegant model is shown in Fig 1.61 above. Exercise 15 (page 54) describes a construction for a simpler device.

There is a huge amount of literature on the Golden section, or Golden proportion, or Golden ratio or Golden cut.

There is even a magazine, the *Fibonacci Quarterly* from the United States which embraces this territory! And it is something that the students at this time need to become aware of through the fact that it has such a ubiquitous appearance in nature (including in the human being). So what is it? It is a ratio, and a very special ratio. It is also called $\varphi$ (*phi*).

## Phi, *Fibonacci and the 'Golden cut' ....*

What is $\varphi$? This is the symbol given to this quite special number. As a crude approximation it is 1.6180 to four decimal places. It appears in numerous places in nature, in many more than we can dream of, but it is also claimed to be in many places where it is dubious. An amazing set of relationships that applies to very many things should *not* be assumed to apply to everything!

## *Exact value for* phi, *calculated from its definition*

The definition is crucial. One definition is the number derived when a line is cut at a particular place such that:

> *The ratio of the whole to the larger part is the same as the larger part to the smaller.*

Graphically this can be shown along a line as:

|   |
|---|
| the whole |

| the smaller part | the larger part |
|---|---|

And the ratios *graphically* expressed would look like this:

Algebraically the exact ratios can be found for this particular division. If we let 1 unit be the length of smaller part and the larger part be unit $x$ units, then the whole will be $1 + x$ units in length. Or again more graphically:

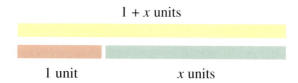

1 + x units

1 unit          x units

And in conventional mathematical symbolism this is shown as in the pannel opposite and we solve for $x$:

$$\frac{x}{1} = \frac{1+x}{x}$$

*Golden ratio*

The Golden ratio or section is ($\sqrt{5}$ + 1)/2 = 1.618033989...
as far as most school calculators will go. To calculate this
exactly we do the following: Given that we want that the
ratio of the whole to the larger is the same as the ratio of
the larger to the smaller then we can write

$$\frac{x}{1} = \frac{1+x}{x}$$

Firstly cross multiply:

$$x^2 = 1 + x$$

Rearrange and equate to zero:

$$x^2 - x - 1 = 0$$

This is now a quadratic equation and it can be solved with
the formula for $x$ as the unknown (not proved for the students until much later):

$$x = \frac{-b \pm \sqrt{b^2 - 4ac}}{2a}$$

where: $ax^2 + bx + c = 0$

and in this case $a = 1$, $b = -1$ and $c = -1$. Hence, substituting for $a$, $b$ and $c$ we get:

$$x = \frac{-(-1) \pm \sqrt{(-1)^2 - 4 \times 1 \times -1}}{2 \times 1}$$

Simplifying gives: $x = \dfrac{1 \pm \sqrt{1 + 4}}{2}$

Or: $x = \dfrac{1 \pm \sqrt{5}}{2}$

So $x$ is: $x = \dfrac{1 + \sqrt{5}}{2}$ or $x = \dfrac{1 - \sqrt{5}}{2}$ exactly.

Yes, that's right, *two* answers. We take the *positive* value
only, rewriting it as:

$$x = \frac{\sqrt{5} + 1}{2}$$

The result from the pannel gives us the usual *exact* value for $\varphi$. Thus we can write:

$$\varphi = \frac{\sqrt{5}+1}{2}$$

And this, as a non repeating decimal is:

$$\varphi = 1.618033988749894848204586834365 64$$

and which is usually remembered as:

$$\varphi = 1.618$$

The negative value $(1-\sqrt{5})/2$ gives the reciprocal $-0.618$. As the number is a ratio, it is immaterial whether expressed as positive or negative.

## 1.618 or 0.618?

This is the number usually designated as $\varphi$ or *phi*. Sometimes its *reciprocal* is also named *phi*. This is somewhat confusing! But most writers seem to use the value 1.618 (Livio 2002, 80; Critchlow 1969, 30; *www.ams.org,* 2004) rather than 0.618 (Ghyka 1946, 7). Some various references are shown in the bibliography.

### Exercise 15    Making a set of Golden section dividers

A simple set of such dividers can be made from card and pins. The basic structure is straightforward. Suitable model dimensions are shown in Fig 1.62.

1. Draw the four card profiles shown (or design your own, or let the students design them).

2. Score along the red dotted lines (this has the effect of stiffening the dividers arms if bent along the dots).

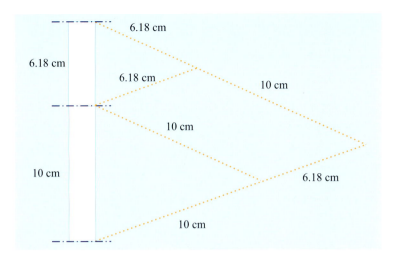

*Fig 1.62  Typical dimensions for a Golden section set of dividers*

3. Pierce holes at the points marked.

4. Cut out the four profiles shown in heavy black outline.

5. Assemble with split pins as pivots.

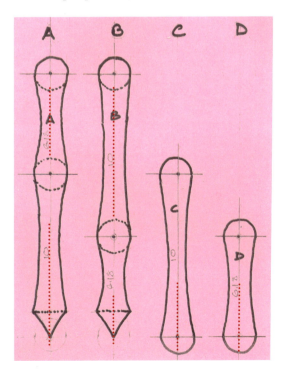

*Fig 1.63*

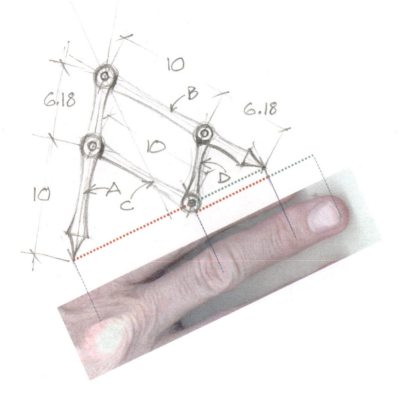

*Fig 1.64  Dividers in use*

6. Check how accurate your dividers are on a ruled measure (10 cm
   with the wider dividers should give about 6.2 cm with the closer
   distance)

We can, for instance, check the proportions of the joints of our
fingers. See how close these are to this Golden section. If we
could see the actual bone joints perhaps it would be more accurate?

Where do we see such proportions in the human head? Is
'MAN' the measure of all things perhaps? Or is this just a mere
coincidence? Such is not my conviction.

*Exercise 16   Golden ratio in the human form*

The students should measure these heights, *A,* and *B,* for themselves to see how close *they* are to this Golden ratio.

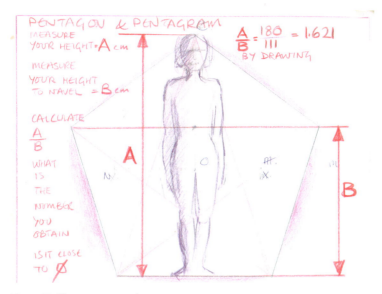

*Fig 1.65  Human proportions*

The young person of this age, not having matured fully, will be somewhat different. See if the group can decide for a greater *or* smaller ratio than $\varphi$ — and *why*.

1. Measure overall height. Let this be *A*.

2. Measure distance to navel from the ground. Let this be *B*

3. Divide *A* by *B*.

4. Make a table of these values.

| A | B | A / B |
|---|---|---|
|   |   |   |
|   |   |   |
|   |   |   |
|   |   |   |
|   |   |   |
|   |   |   |

*Fig 1.66  Table of values*

5. Calculate the average of these ratios. What number is this close to?

The French architect, Le Corbusier, attempted to design a scale which he called the 'Modulor,' based on this proportion, as an aid to design in building. A sketch of this is shown in Fig 1.67.

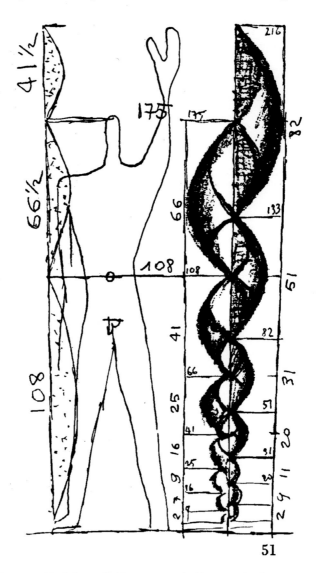

*Fig 1.67  Modulor, a Golden ratio scale (from Le Corbusier, 1954, 51)*

The principal dimensions are 108 and 66.5 in the drawing. If these are added and then divided by 108 we get 1.616, which is close to $\varphi$.

*Exercise 17   Fibonacci in the plant*

Some plants show this Fibonacci series in how they branch. The sneezewort (*Achillea ptarmica*), a type of yarrow, displays this number series in the plane, according to Huntley, as it develops towards its flower heads (Huntley 1970, 163).

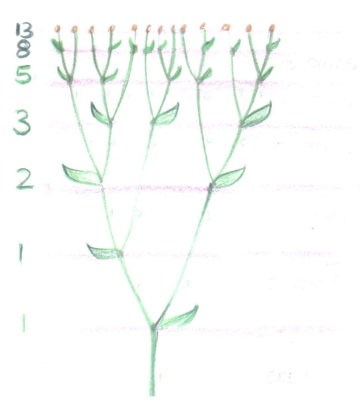

*Fig 1.68  Sneezewort branching*

Note how the leaf nodes appear to line up in this sketch. Does Fennel do something similar? One would have to find some and look! Now that would be an exercise.

*Exercise 18   Phi — The Golden section*

As seen earlier, if we look carefully, there is this particular number which is gradually approached if we take consecutive

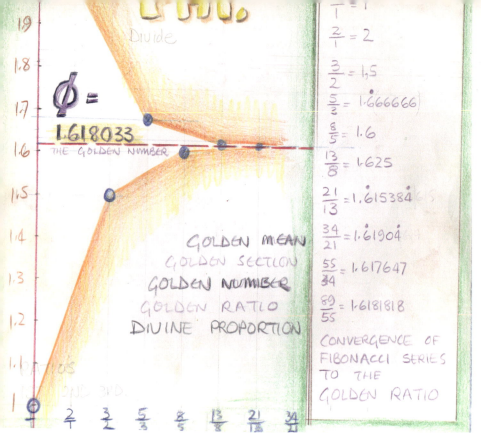

Fig 1.69 *Phi, as a converging limit*

numbers of the Fibonacci series, that is by dividing the latter by the former — for example 5/3 = 1.6666... , and 8/5 = 1.6, and 13/8 = 1.625, etc. These quotients alternate *about* this number but get slowly closer to it — see Fig 1.69.

*Exercise 19   Collage of natures forms*

To conclude the main lesson it may be a good idea to get the students to make a collage of some kind, bringing together drawings, photographs or actual artifacts that exhibit these simple forms — be they galaxies, operculums, pineapples, hen's eggs, tornadoes, seashells, flower buds or a legion of forms their own observation will discover. Some suggestions are in Fig 1.70 and Fig 1.71.

I always hope that students (and adults!) will slowly begin to see where there is some kind of pattern in nature. If there is a pattern then there is sure to be some maths and geometry. Even at this young age the eye can be open to form.

▲ *Fig 1.70  Spiral shapes*     ▼ *Fig 1.71  Eggs, urns, bud and tree forms …*

# 2. Pythagoras and Numbers

This is a main lesson we have often done with this age group of around 12 to 13 (Year 7). It introduces qualities, kinds and notions of *number* in particular. There is always a relation between number and geometry but here number is emphasized whereas in the 'Maths in Nature' section geometry was emphasized.

## Why Pythagoras?

We owe a debt to this time of Ancient Greece, for the individuality of Pythagoras must have been one of many who took such a searching stance towards knowledge that most of us can still be challenged by it and are far from emulating it.

These great individuals saw the whole as a whole, and could also see some of what made it a whole. This was what gave it the order it had. And this was *number.* In the concordance and harmony that spoke through music Pythagoras sought to see how this met the human soul and the way it could lead its life. It is said that in his school, it was music that led the community into the day with special harmonies. Students need to know of his life.

Perhaps that is what these greats were sent for. Thales (*c.* 625–547 BC), Euclid (dates uncertain but *c.* 300–260 BC), Archimedes (*c.* 287–212 BC) and Plato (*c.* 428–348 BC) were among them. And their work lived for around two thousand years before a new approach emerged with the Renaissance in Europe, prepared with the earlier help of Arab scholars. We have to thank much that came from the Arab world for translations and developments which later provided material debate among Renaissance thinkers.

With the students of this age we are at a time when there appears to be a seeking in the young person that asks for a correspondence between what is '*in* the head' and what is *out*side. This is a reflection of what we as humanity went through, and

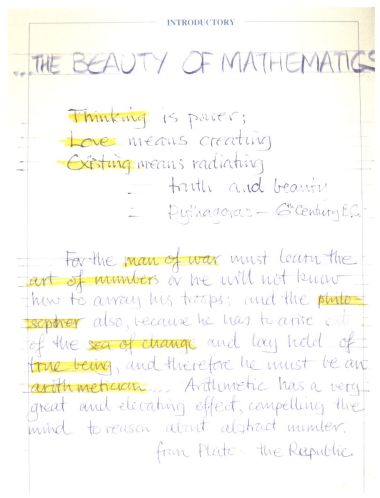

*Fig 2.1 Introductory themes ...*

are still going through historically, with the empiricist and rationalist dichotomy that emerged starkly. It is not so much the dichotomy that is important but the resolution of it. We would not ask a single question if we did not need a resolution of it!

This double-sidedness emerges in this questioning we have. If the inner and outer correspondences were immediately apparent there would be no questioning. But there is. So they are not. And at one age or another this becomes apparent to the young person. We bring a way to meet this emerging problem in such a main lesson as this.

## Number

The two aspects concentrated on in our Year 7 main lessons are number and geometry. *Geometry* was covered in the previous main lesson and *number* is studied in this main lesson of about three weeks. Needless to say these two directions overlap but here there is an emphasis on number relations of all sorts. And in particular this is the history of number and number systems. Why number? Is it that in this world we quickly start to see some order, some patterns?

Not all cultures were obsessed with the number ten, and thus the decimal system, as we are almost universally today. We could of course put this down to the fact of ten fingers, but the ancients in Chaldea had sixty in mind — one has to wonder why.

Or there is *one, two* and *many,* which we could accuse the so-called primitives of working with. One aboriginal group counted with a sort of trinitarian base (Fig 2.3). Here the new repeating step began after three steps, not ten.

Yet there are mysteries in numbers, they are not merely a scheme for counting out the dollars, bums on seats, number puzzles, even the current Sudoku craze, or how high a high-rise measures. There is music here — as well as counting and measuring.

*Fig 2.2 A handy start to counting*

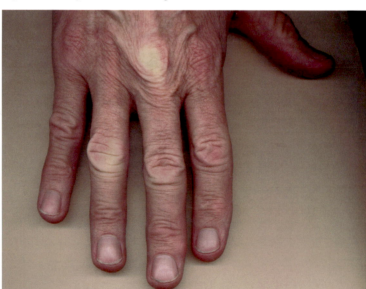

# NUMBER BASES — OR RADIX

Base 10 — or Decimal System, uses ten sym...

This is the counting system we are familiar wi...
for when ten, 10, is reached we start again unt...
the same symbols.

For example:

6735 is shorthand for:

6000 + 700 + 30 + 5

or 6×1000 + 7×100 + 3×10 + 5×1

or 6×10³ + 7×10² + 3×10¹ + 5×10⁰
which is the longhand version!

From an aboriginal dialect

One: mal (1)          Four: bular bular

Two: bular (2)        Five: bular gulibar

Three: guliba (3)     Six: guliba guliba

This system has          guliba guliba
                         6 «35
elements of a          bular guliba
                         5 or 2 3
three fold             bular bular
                         4 or 2 2
base ...     guliba           ...     as shown
                         the step diagram
Bular 2

mal

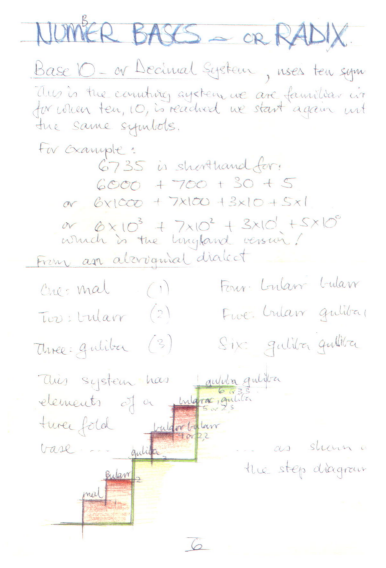

6

*Fig 2.3  An aboriginal dialect — with a threefold base*

## Qualitative number

Number in the abstract is our most common usage. It can be
reduced to mere *calculation*. How much money is in the bank for
instance — as one lady said to me recently when we talked about
maths! But number can be viewed quite differently.

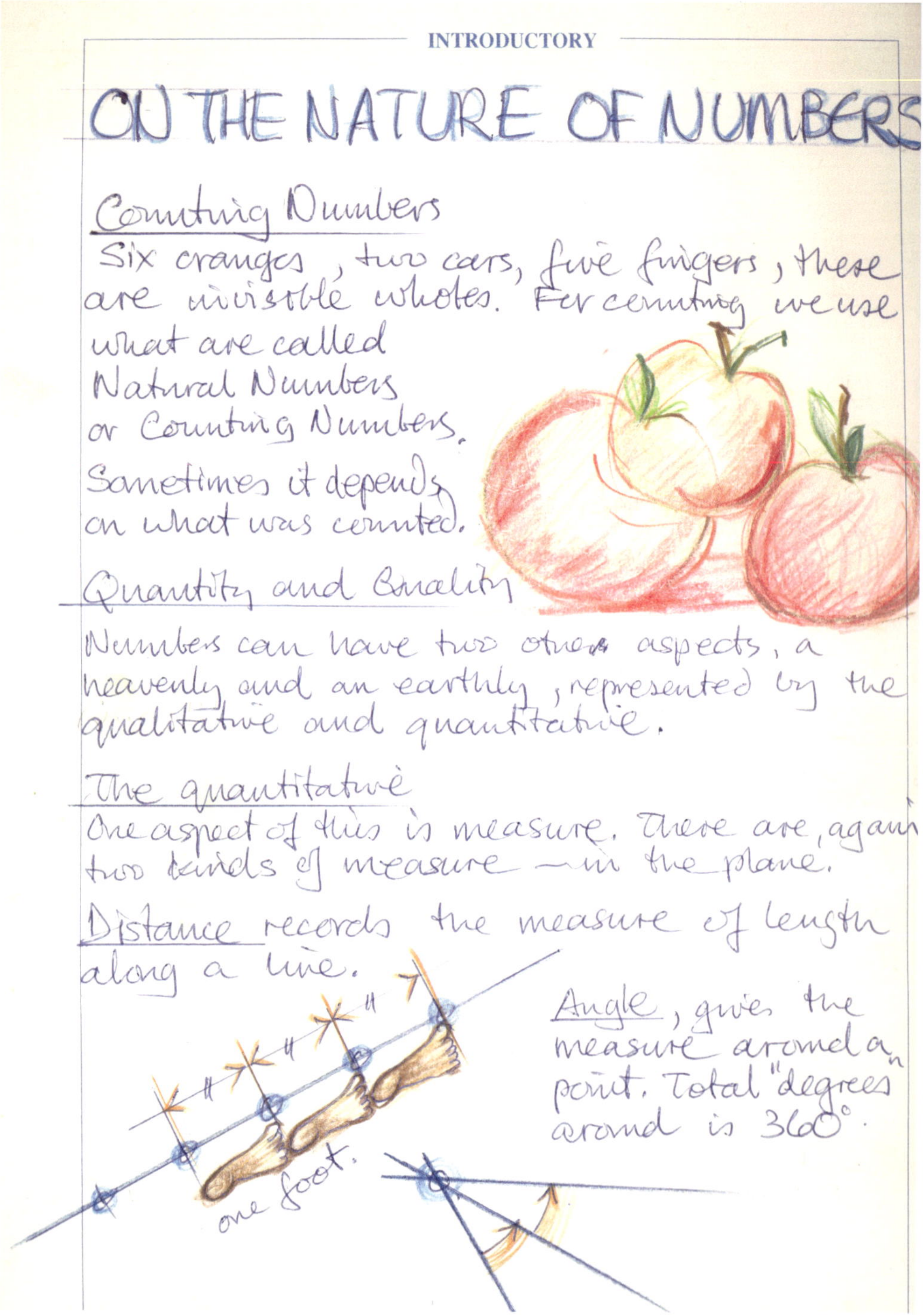

**INTRODUCTORY**

# ON THE NATURE OF NUMBERS

## Counting Numbers

Six oranges, two cars, five fingers, these are invisible wholes. For counting we use what are called
Natural Numbers
or Counting Numbers.

Sometimes it depends on what was counted.

## Quantity and Quality

Numbers can have two other aspects, a heavenly and an earthly, represented by the qualitative and quantitative.

## The quantitative

One aspect of this is measure. There are, again two kinds of measure — in the plane.

## Distance records the measure of length along a line.

Angle, gives the measure around a point. Total "degrees around is $360°$.

one foot.

Fig 2.4  *The qualitative and the quantitative*

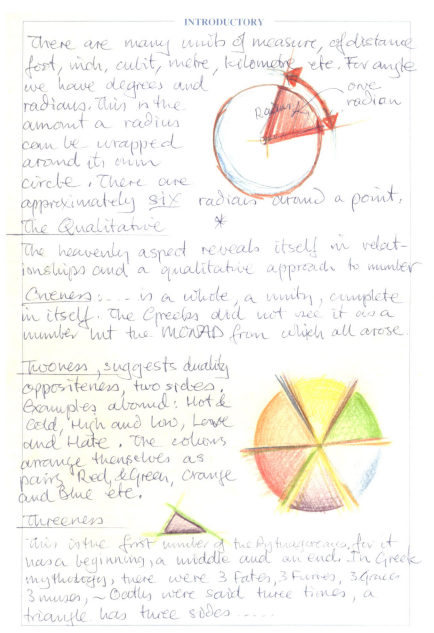

There are many units of measure, of distance foot, inch, cubit, metre, kilometre etc. For angle we have degrees and radians. This is the amount a radius can be wrapped around its own circle. There are approximately six radians around a point.

one radian

Radian

The Qualitative                    *

The heavenly aspect reveals itself in relationships and a qualitative approach to number

Oneness: .... is a whole, a unity, complete in itself. The Greeks did not see it as a number but the MONAD from which all arose.

Twoness, suggests duality oppositeness, two sides. Examples abound: Hot & cold, High and Low, Love and Hate. The colours arrange themselves as pairs Red & Green, Orange and Blue etc.

Threeness
this is the first number of the Pythagoreans, for it has a beginning, a middle and an end. In Greek mythology, there were 3 Fates, 3 Furies, 3 graces 3 muses, ~ Oaths were said three times, a triangle has three sides .....

*Fig 2.5 The heavenly or the qualitative*

Is there something about unity that is essentially *different* from a twoness or duality, or a threefoldness (trinities), sevens or fives or even tens? There is far, far more to it than *the One* and *the Many!*

*Exercise 20*

Explore qualities of the numbers up to at least twelve (say). What can we say about 'one-ness' for instance? Is all one? What about the bits? And so on ... much could be debated.

1. Consider unity.

2. Two. Two-ness. Duality. Dichotomies. Look for opposites. This is where one thing is qualitatively different from another, for instance hot and cold, up and down, plane and point. List at least three more such opposites or dualities.

3. Trinities, can we find any? A fundamental trinity is that in geometry, with the three-some which are the very elements of geometry — that is *point, line* and *plane.* I do not believe it is good enough to say, as Keith Critchlow appears to, that one derives from the other (Critchlow 1976, 10–13) as they are mutually *interdependent* (see Fig 2.6). And are totally different in kind. Are RED, BLUE and YELLOW to be considered a trinity? Most of us will talk of them as the *three* primary colors. Find three more trinities. Dimensions in space ...

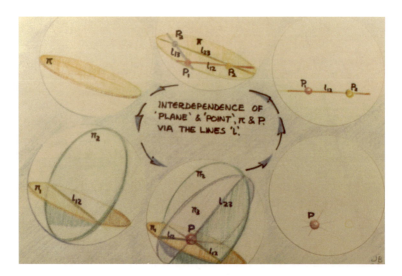

*Fig 2.6 Mutually defining point line and plane. E.g. Three points define a plane. Three planes define a point.*

4. Where do we see the fourfold? Heart beats per breath? Kingdoms of nature?

    Can we find more. Note that when considering the four kingdoms that, although we may say 'four,' each one of them is radically different to another — there is a serious qualitative difference.

5. Where do we see the fivefold? Check out the Rosaceae.

6. Where do we see the sixfold? Check out the Liliacea. some insects — why, one wonders, just *that* many legs? Something to do with three times twofold?

7. Where do we see sevenfoldness?

8. Any 'eights' around? Arachnids.

9. Nines? Christian Hierarchies, rods around a central axis forming the centrioles in cells.

10. Here we can get toey ...

11. This is a hard one ........

12. Dozens?

13. Challenge — to find 'thirteens.'

14. What is essential about 'a hundred'

15. Any other numbers with special qualities?

## Various number systems

Number systems throughout the world have been based on a range of blocks of numbers. From the method of counting in units, to pairs (blocks of two), to threes as mentioned above, to packages of six, to ten (fingers), to twelve (duo-decimal) to twenty and even sixty. And there were many different ways. The

Roman way was different to the Arabic or Hindu way. There have been all kinds of different symbols for each of the numbers. The ancient Egyptians used a base ten — the symbols being as represented in Fig 2.7.

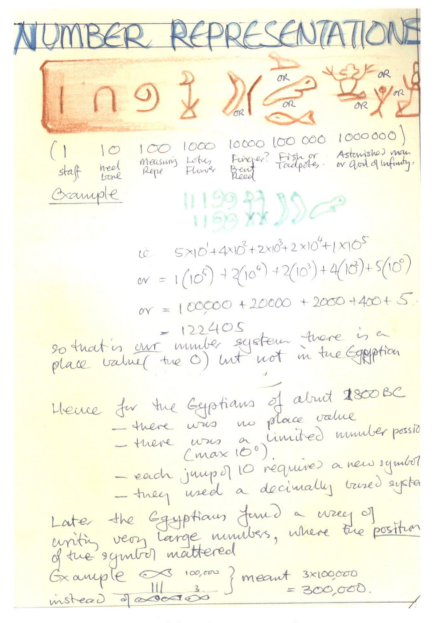

Fig 2.7 Egyptian symbols used to give our number 122405

## *Decimal numbers, whole index form (longhand) and our normal shorthand form*

All of us are familiar with the system of tens. This derives its strength from the notion of place and the inclusion of *nothing, no-thing* or zero. This is called the *base ten* system.

If I write down the number 1 then this is far from 10, which is a lot more than 1 with absolutely *nothing* after it. The fact of putting the zero to the right means that the 1 implies, in this case, *ten* more such ones. So our number depends upon the *place* we put it. To add to the fun we also have different *symbols* for the *number* of 1's we have.

It was the genius of the Hindu and Arab world of yesteryear to give the series of symbols which developed into those we use today. These symbols could have been any kind of glyph but we have ended up with ten of them, if we include the mysterious zero.

These symbols are:

| Symbol | Meaning of symbol |
|--------|-------------------|
| 0 | Nothing |
| 1 | 1 |
| 2 | 1 + 1 |
| 3 | 1 + 1 + 1 |
| 4 | 1 + 1 + 1 + 1 |
| 5 | 1 + 1 + 1 + 1 + 1 |
| 6 | 1 + 1 + 1 + 1 + 1 + 1 |
| 7 | 1 + 1 + 1 + 1 + 1 + 1 + 1 |
| 8 | 1 + 1 + 1 + 1 + 1 + 1 + 1 + 1 |
| 9 | 1 + 1 + 1 + 1 + 1 + 1 + 1 + 1 + 1 |

It may seem trivial to point this out, but **9** is a seriously significant reduction in writing, in time and in space, compared to **1 + 1 + 1 + 1 + 1 + 1 + 1 + 1 + 1**. But there is more. Far greater economies, simplifications and conveniences occur if we write numbers using this place system as well. Having collapsed **1 + 1 + 1 + 1 + 1 + 1 + 1 + 1 + 1** to **9** and not even having reached **10** it is most impressive that we can further collapse large numbers by a further simple device. This is not a symbol as such but a way in which a process is presented.

Recapping all the units ....

1 + 1 + 1 + 1 + 1 + 1 + 1 + 1 + 1 + 1 + 1 + 1 + 1 + 1 + 1 + 1 + 1 + 1 + 1 + 1 +
1 + 1 + 1 + 1 + 1 + 1 + 1 + 1 + 1 + 1 + 1 + 1 + 1 + 1 + 1 + 1 + 1 + 1 + 1 + 1 +
1 + 1 + 1 + 1 + 1 + 1 + 1 + 1 + 1 + 1 + 1 + 1 + 1 + 1 + 1 + 1 + 1 + 1 + 1 + 1 +
1 + 1 + 1 + 1 + 1 + 1 + 1 + 1 + 1 + 1 + 1 + 1 + 1 + 1 + 1 + 1 + 1 + 1 + 1 + 1 +
1 + 1 + 1 + 1 + 1 + 1 + 1 + 1 + 1 + 1 + 1 + 1 + 1 + 1 + 1 + 1 + 1 + 1 + 1 + 1

Collecting these in blocks of 10
= 10 + 10 + 10 + 10 + 10 + 10 + 10 + 10 + 10 + 10

Multiplying the number *of* blocks *by* the number *in* each block.
= 10 × 10

Which is:
= 100

And now if we raise the block to the *power* of the number of blocks, i.e. 2 in this case, we can write $10^2$. This is the ultimate condensation (so to say) used today.
   So we can be extra economical with large numbers.

| | | | |
|---|---|---|---|
| 100 | = | 10 × 10 | = $10^2$ |
| 1 000 | = | 10 × 10 × 10 | = $10^3$ |
| 10 000 | = | 10 × 10 × 10 × 10 | = $10^4$ |
| 100 000 | = | 10 × 10 × 10 × 10 × 10 | = $10^5$ |

And even *very* large numbers. Would any one like to write out 10 000 000 000 000 000 000 000 000 every time? No! So we write, simply, $10^{25}$ which means, ... well go figure!

## Longhand and shorthand

This method enables even these index or power numbers to be written very simply. If we have the number 76 540 (say) then what this really means is:

the sum of:
  70 000 + 6 000 + 500 + 40

or the sum in multiples of the tens
  7 × 10 000 + 6 × 1000 + 5 × 100 + 4 × 10

which is when written as powers
  $7 \times 10^4 + 6 \times 10^3 + 5 \times 10^2 + 4 \times 10^1$

So the *shorthand* version, 76540
  *means,* in longhand $7 \times 10^4 + 6 \times 10^3 + 5 \times 10^2 + 4 \times 10^1$

*Exercise 21 — Decimals in long and shorthand*

1. Write out 360 in *longhand*
   $3 \times 10^2 + 6 \times 10^1$

2. Write in shorthand   $5 \times 10^4 + 9 \times 10^3 + 1 \times 10^2 + 2 \times 10^1$
   59 120

3. So longhand for 365 is?
   $3 \times 10^2 + 6 \times 10^1 + 5 \times 10^0$

4. What then is
   $5 \times 10^4 + 3 \times 10^3 + 2 \times 10^2 + 9 \times 10^1 + 7 \times 10^0$
   in shorthand?
   53 297

(Note: we have implied that $10^0 = 1$. To justify this may be a bit much for some Year 7's but if they can get the hang of indices or powers then they could be led through the following:

$$1 = \frac{5}{5} = \frac{5^1}{5^1} = 5^{1-1} = 5^0$$

(This means that if all these equalities *are* equal then $5^0 = 1$. But we could have chosen *any* number, not just 5. Hence we can say that $x^0 = 1$, where $x$ is any number. So that $10^0 = 1$ if $10 = x$. In other words anything to the power zero is one.

Now that means also that $2^0 = 1$ as well. We will need this to continue to deal with binary numbers — in other words, using base 2, rather than base 10.

## *Binary numbers*

A modern number system that we find in the calculator and computer is that based on the number two, or *base two* system. The students may not come across this much yet — or will they? Base two is interpreted often as *on* or *off*, two key conditions of any electrical circuit. *Active* or *Inactive*. And these are often represented symbolically as **1** (active) and **0** (inactive).

This figure **1** means *on* and **0** usually means *off*. Sometimes control switches on electrical hardware have markings just so although mostly we see a kind of combination of a **1** and an **O**.

Sometimes we also see numbers of the form **101011**. What does this mean? How big, or small, is this number?

If all these ones and zeros represent the presence or absence of the number two to some power, and place is also important how then do we interpret such numbers? Just as we considered powers of **10** for base **10** numbers, we consider powers of **2** for base two numbers.

It is helpful to make a table of powers of two to start with. The second column becomes familiar with usage.

*Fig 2.8  Machine on / off switch combining*
*1 and 0 in its marking*

| | |
|---|---|
| $2^0 = 1$ | 1 |
| $2^1 = 2$ | 2 |
| $2^2 = 2 \times 2$ | 4 |
| $2^3 = 2 \times 2 \times 2$ | 8 |
| $2^4 = 2 \times 2 \times 2 \times 2$ | 16 |
| $2^5 = 2 \times 2 \times 2 \times 2 \times 2$ | 32 |
| $2^6 = 2 \times 2 \times 2 \times 2 \times 2 \times 2$ | 64 |
| $2^7 = 2 \times 2 \times 2 \times 2 \times 2 \times 2 \times 2$ | 128 |
| $2^8 = 2 \times 2 \times 2 \times 2 \times 2 \times 2 \times 2 \times 2$ | 256 |
| $2^9 = 2 \times 2 \times 2 \times 2 \times 2 \times 2 \times 2 \times 2 \times 2$ | 512 |

*Exercise 22   Converting binary to base 10*

1. What is the binary number     **1**         **1**            **1**   to base 10?

| | | |
|---|---|---|
| This means | = | $1 \times 2^2 + 1 \times 2^1 + 1 \times 2^0$ |
| or | = | $1 \times 4 + 1 \times 2 + 1 \times 1$ |
| or | = | $4 \quad + 2 \quad + 1$ |
| or that is | = | 7 |

This is sometimes written as   $111_2 = 7_{10}$

2. Another example. What is     **1**       **0**       **1**       **0**
   to base 10?

| | | |
|---|---|---|
| | = | $1 \times 2^3 + 0 \times 2^2 + 1 \times 2^1 + 0 \times 2^0$ |
| | = | $1 \times 8 + 0 \times 4 + 1 \times 2 + 0 \times 1$ |
| | = | $8 \quad + 0 + 2 + 0$ |
| | = | 10 |
| That is | | $1010_2 = 10_{10}.$ |

3. What is **1000₂** converted to base 10?          8

4. What is **101011₂** converted to base 10?  32+0+8+0+2+1 = 49

5. But what about **145₁₀** converted to binary or base 2? (see table above)
   First subtract the highest power of 2 less than 145.
   That is 145 – 128 = 17.
   Now subtract the highest power less than 17.

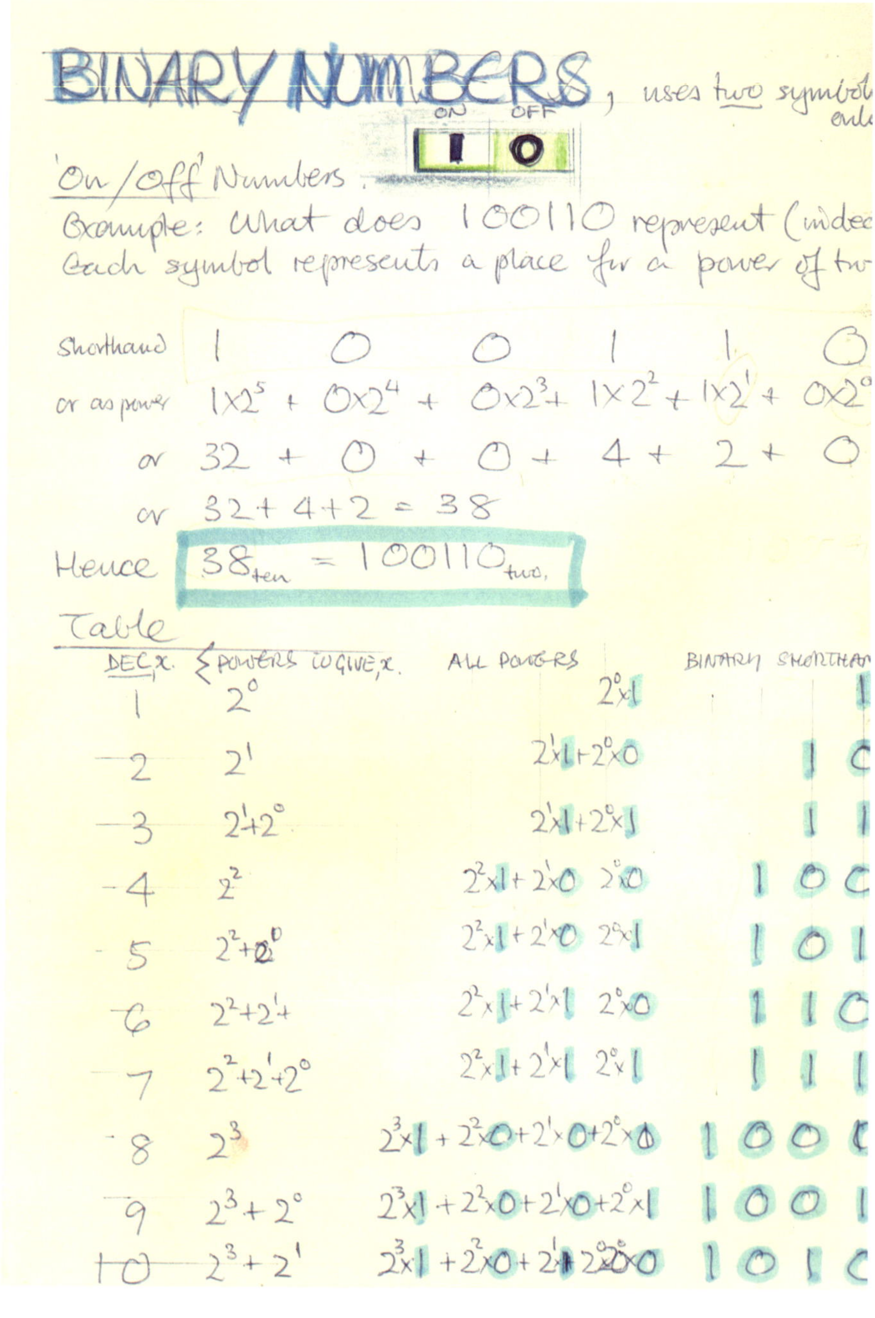

# BINARY NUMBERS, uses two symbols only

ON   OFF

| I | O |

'On/Off' Numbers.

Example: what does 100110 represent (indec

Each symbol represents a place for a power of two

| Shorthand | I | O | O | I | I | O |
|---|---|---|---|---|---|---|
| or as power | $1 \times 2^5$ + | $0 \times 2^4$ + | $0 \times 2^3$ + | $1 \times 2^2$ + | $1 \times 2^1$ + | $0 \times 2^0$ |
| or | 32 + | O + | O + | 4 + | 2 + | O |
| or | 32 + 4 + 2 = 38 | | | | | |

Hence  $38_{ten} = 100110_{two}$,

Table

| DEC x. | $\{$ POWERS TO GIVE x. | ALL POWERS | BINARY SHORTHAND |
|---|---|---|---|
| 1 | $2^0$ | $2^0 \times 1$ | I |
| 2 | $2^1$ | $2^1 \times 1 + 2^0 \times 0$ | I O |
| 3 | $2 + 2^0$ | $2^1 \times 1 + 2^0 \times 1$ | I I |
| 4 | $2^2$ | $2^2 \times 1 + 2^1 \times 0 \ 2^0 \times 0$ | I O O |
| 5 | $2^2 + 2^0$ | $2^2 \times 1 + 2^1 \times 0 \ 2^0 \times 1$ | I O I |
| 6 | $2^2 + 2^1$ | $2^2 \times 1 + 2^1 \times 1 \ 2^0 \times 0$ | I I O |
| 7 | $2^2 + 2^1 + 2^0$ | $2^2 \times 1 + 2^1 \times 1 \ 2^0 \times 1$ | I I I |
| 8 | $2^3$ | $2^3 \times 1 + 2^2 \times 0 + 2^1 \times 0 + 2^0 \times 0$ | I O O O |
| 9 | $2^3 + 2^0$ | $2^3 \times 1 + 2^2 \times 0 + 2^1 \times 0 + 2^0 \times 1$ | I O O I |
| 10 | $2^3 + 2^1$ | $2^3 \times 1 + 2^2 \times 0 + 2^1 \times 1 \ 2^0 \times 0$ | I O I O |

Fig 2.9  Binary numbers

That is $17 - 16 = 1$.
Thus we have $145 = 128 + 16 + 1$.
Or, putting in missing powers:

$145 = 1{\times}128 + 0{\times}64 + 0{\times}32 + 1{\times}16 + 0{\times}8 + 0{\times}4 + 0{\times}2 + 1{\times}1$
$145_{10} = 1 \quad\quad 0 \quad\quad 0 \quad\quad 1 \quad\quad 0 \quad\quad 0 \quad\quad 0 \quad\quad 1_2$

6. What is $999_{10}$ in binary?

$512 + 256 + 128 + 64 + 32 + 0 + 0 + 4 + 2 + 1 = 1111100111$

7. And finally show that $38_{10}$ is the same as $100110_2$.

## Measurement

Counting numbers and their systems are one thing but there is something else to count apart from things, for measurement is a thing in itself! If we wish to *measure* anything too we have recourse to number — but also something else. To count distance, for instance, we have to have some standardized measure. The different standards of measure are legion. Or were. Today many countries use the metric system, others the imperial (even the USA). Before that there were many other standards. Before going too far we need to recognize there are two fundamentally different kinds of measure in the plane.

## Distance and angle

In our ordinary day-to-day space these two are *distance* and *angle*. That is distance *along* a line and angle *around* a line (or point). Imagine how different these are.

In the one case we step along a line in, let us say, equal lengths. A good measure of such distances would be our length of pace, or for somewhat shorter, the length of our foot. This is indeed a measure — the foot — still used in many places.

In early times another measure was used by a number of cultures, the *cubit*. But how long was a cubit? Was this the length of

# UNITS OF MEASURE

Why do we measure? To compare of thing with another — for length, for size, for area for volume and for length of time.

Early in human history artefacts were compared with the human form. The human was the standard. Much later part of the earth's surface became the basi for measure (the metre). Later still the wavelength of a coloured light.

Babylonian and Egyptian peoples, long ago used the CUBIT, ie the length of the forearm from elbow to finger tips.

The Cubit.

Comparison of Students

Historically: Egypt 52.3 cm Babylon 49.61 cm. Assyria 55.37cm Asia Minor 51.74 cm.

Fig 2.10  Kinds of measure: the cubit

Related to the human body are:

Digit —— width of forefinger, 19mm

Hand —— standardised at 4" or 10.2cm.

Span —— about 24cm.

Pace —— about ..

Fathom —— now 6'.00 or 183cm

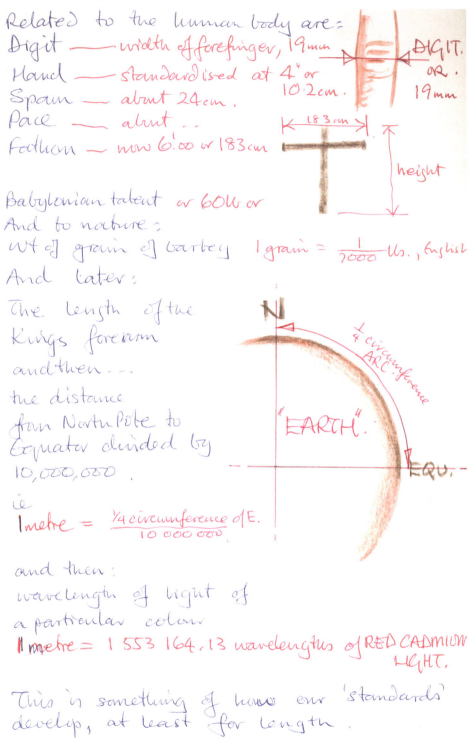

Babylonian talent or 60lb or

And to nature;

wt of grain of barley   1 grain = $\frac{1}{7000}$ lbs., English

And later:

The length of the King's forearm and then ...

the distance from North Pole to Equator divided by 10,000,000.

ie

$$1\text{metre} = \frac{\frac{1}{4}\text{ circumference of E.}}{10\ 000\ 000}$$

and then:

wavelength of light of a particular colour

1 metre = 1 553 164.13 wavelengths of RED CADMIUM LIGHT.

This is something of how our 'standards' develop, at least for length.

*Fig 2.11  From the King's forearm to wavelengths of light*

the forearm of the Pharaoh or the King? If so, then it inevitably varied. Even among different peoples it varied. For the Egyptians and Babylonians it was about 20 inches or 51 cm (Oxford Junior Encyclopedia 1951, 263). Other sources give other values (see Fig 2.10).

The historical progression of measures seems to have been taken from parts of the highest of the human in the hierarchy (king, etc.), the human form as such, thence to nature (a portion of the Earth's circumference), and, of late, to a certain number of the wavelengths of light.

One quarter the distance from the North Pole to the equator divided by 10 000 000 was deemed in about 1791 by the French Academy of Sciences to be the metre. The epic efforts of Méchain and Delambre resulted in the value of 39.37008 inches for the metre. The word derives from the Greek word that means measure, *metron*. Nowadays there is a reverse definition in that the metre is said to be that distance traveled by light in a vacuum in 1/299 792 458 of a second! (This assumes that light has a 'speed' in the ordinary sense — which is contestable.) This is given in the SI (International System of Units) definition. Most of us are happy to pace something out if we want a reasonable idea of the length of a housing block. But for some purposes much greater accuracy is needed. Especially if you are buying the block!

*Exercise 23   Distance measures*

1. Find a number of *distance* measures and state their estimated lengths in both inches and centimetres in a table.

2. How long is one inch in millimetres?   25.4

3. Find the average cubit length in Fig 2.10 above.   52.255 cm

4. Why might the initial French definition of the metre be questionable?

   It assumes that the Earth is an exact sphere. It is not. It is at least a geoid, pear shaped and tetrahedral all at the same time.

5. What was an 'inch' based on?

A statute from 1284 gives 'Three grains of barley, dry and round, make an inch; twelve inches make a foot; three feet make an ulna [yard].'

## Angular measure

This is a different world. Now there are far fewer units that appear to be used. I have only discovered three. The three are *degree, radian* and *gon* or *grad.* All are based on the full revolution of a circle. The degree is by far the most common. Radians are more for the mathematical types. And the 'gon' I only came across this year.

| Unit: | Portion of a full circle: |
|---|---|
| Degree | 360 to circle |

*(Each degree is made up of 60 minutes and each minute is of 60 seconds)*

| | |
|---|---|
| Radian | $1/2\pi$ (or about 57 degrees). About 6.28 to circle |

*(We deal with some of the mysteries of $\pi$ later in these notes)*

| | |
|---|---|
| Gon, grad | 400 to circle |

*(Each gon or grad is made up of 100 centesimal minutes and each minute is of 100 centesimal seconds)*

The gon is used by surveyors in some European countries. It is an attempt to decimalize a portion of the circle, namely the right-angle. So there are 100 gons to the right-angle. I was familiar with this as the *grad.* If we examine the school calculator we find a choice of *Deg, Rad* or *Gra,* where Gra implies grad. Check it out on your school calculator.

*Exercise 24   Angular measure*

1. How many degrees in a quarter of a circle?   360/4 = 90

2. How many degrees in 100 revolutions?

$$360 \times 100 = 36\ 000$$

3. How many gons in 8.345 cycles around a centre?

$$8.345 \times 400 = 3338$$

4. If $\pi$ = 3.141592653589793 radians then how many radians in a full circle?        3.141592653589793 × 2 = 6.2831853071796

5. How do each of these three compare, that is if a degree is *one* unit how big or small is a grad and how big is a radian in degrees
   1 : ? : ?

   A grad is 90/100 degrees = 0.9 degrees. A radian is 360 / $2\pi$ degrees = 59.2957 degrees. Hence   1 : 0.9  : 59.2957

## Familiar measuring instruments

Distance is measured with a straight ruler graduated in centimetres (as well as in inches in some countries) and usually 30 cm long.

Angle is measured with a protractor usually graduated in blocks of ten degrees in 1 degree divisions and takes the form of a semi-circle (less commonly a full circle).

These two instruments indicate two fundamentally different worlds — linear (straight line) and the circular (the curve). The distinction between the two has significant ramifications later on, as the measures of one are *incommensurable* with the other in whole-number terms. We see this in the circle where for a circle of 2 units diameter the circumference is

2 × $\pi$ = 2 × 3.141592653589793 ...
= 6.2831853071796 .... units.

This is connected with one of the three famous ancient problems, the one framed as 'squaring the circle,' where it was put out as a challenge to find with compass and ruler how to construct a square equal in its area to a circles area of known radius.

*Fig 2.12 The two basic drawing instruments*

This was in due time proved to be impossible and could not have been solved by the ancients as it requires the length of $\sqrt{\pi}$ 'but classical constructions can only produce algebraic numbers' (Gullberg 1997, 422) and $\pi$ is a transcendental number, not just an irrational.

## Kinds of numbers

The various kinds of numbers are summarized below in Fig 2.13 (so far as the real numbers are concerned.)

## Prime numbers and the sieve of Eratosthenes

There is a whole fascinating range of numbers greater than one for which we can find nothing which will divide into them (except the number itself — and one). These are called the *prime* numbers. If *one* is not considered prime then the first prime is *two*. The next is three. But four is not prime, as it can be divided by 2.

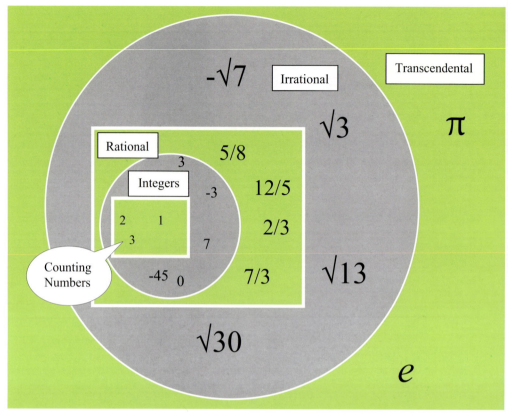

*Fig 2.13 Number sets from counting numbers to transcendental numbers*

They are mysterious as no one can give a definitive predictive law concerning them: a number can only be checked as to *whether* it is a prime.

How is this done? Simply by finding if it has any factors. What is a factor? Leonardo Fibonacci of Pisa called the primes *incomposite* numbers because if they were composite they were not primes. A factor is a number which divides the supposed prime giving another number with no remainder. For example: 99 is not a prime. It can be divided by 9 giving 11 with no remainder.

But 97 *is* a prime. It will not divide by 2 or 3. Or by 4, 5, 6, 7, 8, 9 or 10 and leave no remainder. Nor will it divide by any number between 11 and 98. Test this. It gets tedious after a while. Imagine trying to check whether 987 654 321 is a prime! There are a few initial rules that can help. Is the number 987 654 322 a prime? No, it is not, as any number ending in an even number has at least one factor and that is 2. Suddenly we can eliminate a whole swag of numbers — all the evens or *multiples of two*.

So in the sequence:

1, 2, 3, 4, 5, 6, 7, 8, 9, 10, 11, 12, 13, 14, 15, 16, 17, 18, 19 and 20 we note that all the blue numbers are even, multiples of two and that that is half of them. Does this mean that half the numbers all the way to infinity are even and the other half prime? It may seem so, but is not so at all, as some of these will divide by 3. These are shown in green:

1, 2, 3, 4, 5, 6, 7, 8, 9, 10, 11, 12, 13, 14, 15, 16, 17, 18, 19 and 20 And 5, in red:

1, 2, 3, 4, 5, 6, 7, 8, 9, 10, 11, 12, 13, 14, 15, 16, 17, 18, 19 and 20

This could go on and on and would indeed be tedious but all the numbers *left in the black* would be prime. Not surprisingly there is a diagrammatic or pictorial method worked out by a Greek named Eratosthenes (*c.* 276–194 BC). What he devised is called the *Sieve of Eratosthenes* (Gullberg 1997, 77).

## Sieving for primes

After recognizing that 1 is not considered a prime, we shade alternate numbers *after* 2. Then shade every third number *after* 3 if not already shaded (as in Fig 2.14). And every fifth number *after* 5 if not already shaded and so on. And every seventh after 7 etc. ... Note that the numbers remaining *unshaded* are all prime.

| 1 | 2 | 3 | 4 | 5 | 6 | 7 | 8 | 9 | 10 |
|---|---|---|---|---|---|---|---|---|---|
| 11 | 12 | 13 | 14 | 15 | 16 | 17 | 18 | 19 | 20 |
| 21 | 22 | 23 | 24 | 25 | 26 | 27 | 28 | 29 | 30 |
| 31 | 32 | 33 | 34 | 35 | 36 | 37 | 38 | 39 | 40 |
| 41 | 42 | 43 | 44 | 45 | 46 | 47 | 48 | 49 | 50 |

*Fig 2.14  Eratosthenes sieve from 1 to 50*

*Exercise 25*

1. In the table below continue the shading up to 100. This leaves primes only.

| 1 | 2 | 3 | 4 | 5 | 6 | 7 | 8 | 9 | 10 |
|---|---|---|---|---|---|---|---|---|---|
| 11 | 12 | 13 | 14 | 15 | 16 | 17 | 18 | 19 | 20 |
| 21 | 22 | 23 | 24 | 25 | 26 | 27 | 28 | 29 | 30 |
| 31 | 32 | 33 | 34 | 35 | 36 | 37 | 38 | 39 | 40 |
| 41 | 42 | 43 | 44 | 45 | 46 | 47 | 48 | 49 | 40 |
| 51 | 52 | 53 | 54 | 55 | 56 | 57 | 58 | 59 | 60 |
| 61 | 62 | 63 | 64 | 65 | 66 | 67 | 68 | 69 | 70 |
| 71 | 72 | 73 | 74 | 75 | 76 | 77 | 78 | 79 | 80 |
| 81 | 82 | 83 | 84 | 85 | 86 | 87 | 88 | 89 | 90 |
| 91 | 92 | 93 | 94 | 95 | 96 | 97 | 98 | 99 | |

2. List the primes from 2 to 100

2, 3, 5, 7, 11, 13, 17, 19, 23, 29, 31, 37, 41, 43, 47, 53, 59, 61, 67, 71, 73, 79, 83, 89, 97

3. How many primes are there from 2 to 100?      25

4. How many *new* colors did you have to use?

None as 11, 13, 17 have no multiples not already shaded in

5. Can you detect any patterns in the array of primes?

Primes sometimes come in pairs of numbers with one between. All prime numbers are one more or one less than a number divisible by six.

6. Is $1000011_2$ a prime number? (Note: Primeness is independent of base)

$1000011_2 = 1 \times 64 + 0 \times 32 + 0 \times 16 + 0 \times 8 + 0 \times 4 + 1 \times 2 + 1 \times 1 = 2^6 + 2^1 + 2^0 = 67_{10}$ . Yes.

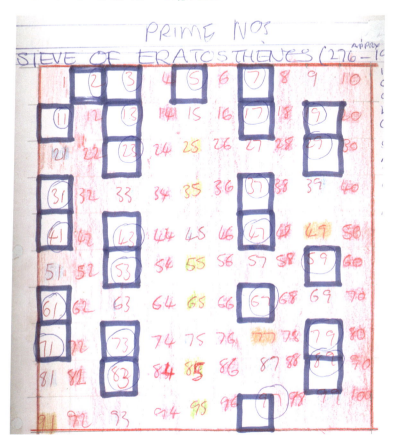

*Fig 2.15  Sieve of Eratosthenes — another presentation*

## Pythagorean Triads

What is the number relationship that Pythagoras himself is so well known for?

### Exercise 26   Pythagorean Triads

This is an exercise to explore what we call 'Pythagorean Triads.' What two whole numbers, which, when their squares are added, will lead to a further *whole* number that is itself a whole number *squared?* (We note that four squared, or $4^2$, means $4 \times 4$.)

Do the following two numbers fit this requirement?

1. 40 and 30? Yes, as $30^2$ added to $40^2$, or $(30 \times 30) + (40 \times 40)$, gives 2500. And 2500 is of course $50^2$, that is 50 is the *square root* of 2500. Now test these.

2. 20 and 30?

3. 3 and 4

4. 5 and 13

5. 12 and 5

6. 1 and 1

7. 240 and 250

Is there a rule that we can find that will tell us how we can set up such a *Pythagorean Triad* — that is not trial and error?

Yes there is, and it is not too difficult and this is elaborated in the next exercise (Exercise 27). But not from all and any numbers that we try, will we get two numbers that will give a *perfect square.*

If, however, our whole number is not a perfect square *can* we find this square root? There is an *algorithm* (or method) and it is calculator free, but more later on this.

Meanwhile how do we form such triads?

*Exercise 27   Pythagorean Triads, a method to determine them*

For the three pro-numerals (i.e., letters representing some unknowns), *a, b* and *c* we let
$$a^2 + b^2 = c^2$$

Now we let     $a = 2pq, \quad b = p^2 - q^2 \quad$ and $\quad c = p^2 + q^2$

where $p$ and $q$ are both positive integers (i.e. positive whole numbers), also where $p$ is *greater* than $q$ (or, symbolically, $p > q > 0$).

With these provisos we can construct some *triads*. Try the following:

1. Let $p = 4$ and $q = 3$ (they are both positive integers, i.e. $4 > 3$, and $p$ is even, but $q$ is not. What are $a$, $b$ and $c$?

So $a = 2 \times 4 \times 3 = 24$, $b = 4^2 - 3^2 = 7$ and $c = 4^2 + 3^2 = 25$

(Check that:  $a^2 + b^2 = c^2$

then substituting $24^2 + 7^2 = 24 \times 24 + 7 \times 7$ where $a = 24$ and $b = 7$

and so                                      $= 576 + 49$

and                                          $= 625$

hence                                        $= 25^2$, and this is $c$ squared as expected)

Now try these ......

2. Let $p = 2$ and $q = 1$  (this is a favorite!)   What are $a$, $b$ and $c$?

3. Let $p = 3$ and $q = 2$  What are $a$, $b$ and $c$?

4. Let $p = 5$ and $q = 2$  What are $a$, $b$ and $c$?

5. Let $p = 4$ and $q = 2$  What are $a$, $b$ and $c$?

6. List the first four Pythagorean Triads.

Are there other conditions which need to be imposed on $p$ and $q$? Books say that one number, but not both, must be even (i.e. divisible by two). Is this so?

The above method can find the 'pure' triads so to say. But it is still possible to find a value for $c$ without special conditions. We can see this graphically (or geometrically) with a further exercise. But first a little exploration with some diagrams.

The answer to No. 2 above was $a = 4$, $b = 3$ and $c = 5$. What does it mean in *space* (or rather the plane) to say that the sum of the square of a number and another different number is equal to the square of a third number (Fig 2.16)?

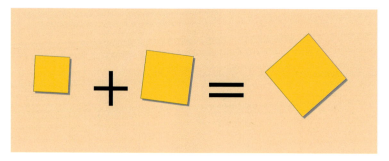

*Fig 2.16  Adding squares?*

Can this be arranged a little more meaningfully? Sure can. We find we can place these squares together to form a triangle *in space* (Fig 2.17). It is a most interesting triangle too. For it has three angles (which, as we know, add to 180°) and one of them is a right angle, or 90° (Fig 2.18). There is a mysterious relationship here between number and geometry ... discovered long ago, it seems, in Sumeria, and credited to Pythagoras.

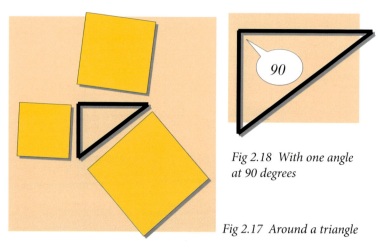

*Fig 2.18  With one angle at 90 degrees*

*Fig 2.17  Around a triangle*

Much more could be said about squares and right angle triangles but for the moment let us draw the *longest* side of a right-angle triangle of unit height and see where this leads.

## Exercise 28 *Finding the longest third side of a specific right-angle triangle*

Starting with *unit* side lengths (i.e. side lengths of *one* unit), for a right-angle triangle how do we find the third side? We can, of course, draw it, and this will give us an answer.

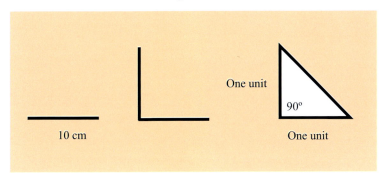

*Fig 2.19*

1. Draw it. That is mark, as a base, a horizontal line 10 cm long (say). We let the unit side be 10 cm long for convenience.

2. Now draw a vertical line from the left-hand end up vertically 10 cm (recall that the construction in Exercise 1 can be used).

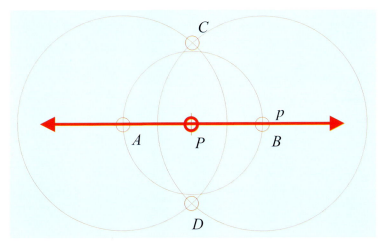

*Fig 2.20*

3. Measure the longest length as accurately as you can. It should be about 14 cm. Or perhaps 14.1 cm. Check. Can we get more accurate still? Is it 14.14 cm maybe? Who would dare to say we can measure to 14.142 cm?

From our numbers we are saying that, as the triangle is right-angled the sum of the squares on the two shorter sides is the same as the square on the longest, the so-called *hypotenuse*.

Now $1^2 + 1^2 = 2$, OK so far.

Or for us here $10^2 + 10^2 = 200$.

But what number *squared* is 2, or, for us, what number squared is 200? Is it 14, or 14.1, or 14.14 or even 14.142? Check the squares of these numbers.

$14 \times 14 = 196$.

$14.1 \times 14.1 = 198.81$. Not too far from 200. And

$14.14 \times 14.14 = 199.9396$. Even closer to 200. And finally (if we *could* measure this accurately) $14.142 \times 14.142 = 199.996164$ which is very much closer to 200.

It could be an interesting exercise in both accuracy of measurement and percentages to see how accurate our actual measures were. So ....

*Exercise 29   What is the square root of 200?*

The students should try to measure first to *no* decimal places (or nearest cm) on their standard centimetre ruler, then to *one* decimal place (i.e. to nearest millimetre) and then to *two* decimal places (at best an estimate to 1/10 of a millimetre). To try to go further with a standard ruler makes little sense as the distances were most likely drawn in the first place with said ruler! Now make a table. As practice complete the table below.

*Fig 2.21*

| Measurements in centimetres | Square this measurement | Subtract from 200 | Divide answer by 200 | and multiply by 100 to give % error |
|---|---|---|---|---|
| 14 | $14 \times 14 = 196$ | $200 - 196 = 4$ | $4/200 = 0.02$ | $0.02 \times 100 = 2\%$ |
| 14.1 | | | | |
| 14.14 | | | | $= 0.0302\%$ |

What do we notice about the accuracy? Does it get better with more decimal places?

It should of course — but will we ever get to exactly 200? I think not. So the students can begin to see we may have some funny numbers here, numbers for which we cannot get an exact square root.

We cannot actually answer what the square root of 200 is but we do know it is *about* 14.14.

The best of the above is 14.14 and even this gives an error of 0.0302%. Which is not bad. But not *exact* either. This simple little 10 by 10 by 14.14 triangle gave the Greek world a lot of trouble, it is said. For here was a triangle which had no *whole* number for one of its lengths, and was not the world constructed from number?

A further exercise can readily find the first few of these funny numbers. These particular funny numbers are called *surds,* (not *absurds*) or *irrational* numbers. The point is that we have found numbers which are not simple counting numbers. And all because of a few triangles, as the following exercise demonstrates.

## *Exercise 30  The first few surds*

Initially make a ten times scale drawing — this is for the sake of accuracy.

1. Mark a point *O* and draw a line to the left 10 cm long to a point *A*.

2. From *A* draw a perpendicular upwards 10 cm to *B*.

3. Join *OB* forming the triangle *OAB*.

4. Measure *OB* and make this the radius of a circle.

5. Draw an arc from *B* down to the line *OA* finding point *C*. *OC* will be about 14.14 cm (or √200)

6. Draw a further perpendicular up from *C* of 10 cm length to *D*.

7. Join *OD* forming triangle *OCD*.

8. Measure *OD* and make this the radius of another circle.

9. Draw an arc from *D* down to the line *OA* finding point *E*. *OC* will be about 17.32 cm (or $\sqrt{300}$)

10. This whole process can be continued indefinitely! But the students should try to get to the next whole number in the sequence. (This will be 3 of course) This is also a good test for accuracy of drawing and compass work.

11. Write in all the radii along the line *OA*. These should be about 10 cm at *A*, 14.14 cm at *C*, 17.32 cm at *E*, 20 cm etc, etc (underneath there are written here only the multiples of the unit values e.g. 1, $\sqrt{2}$, $\sqrt{3}$, 2, $\sqrt{5}$, etc.)

Here we have generated quite a few of these funny numbers. Numbers like 1, 2, 3, 4 and 5 are whole numbers (or integers), while such as $\sqrt{2}$, $\sqrt{3}$ and $\sqrt{5}$ are what we call surds or irrational numbers.

*Fig 2.22*

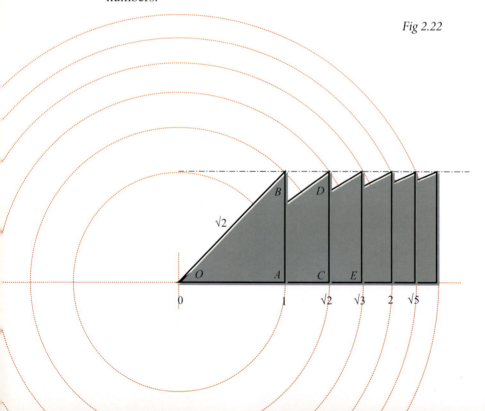

Given that we now have a kind of number which can only be expressed *exactly* using the surd symbol ( √ ) how can we get a good approximation to this — even if not precisely — and without a calculator? For we would not find it helpful to give surds to a manufacturer or carpenter! There is a method (or algorithm) to find a square root. It is a little complicated but not too hard. The first number tested will be a number to which we can work out the answer, simply by ensuring we already know it. For example if we multiply 678 × 678 and get 459 684, then we know that √459 684 is 678. The method is demonstrated in the next exercise. It is a little like a special long division.

## Exercise 31   *Using an algorithm to find a square root*

1. Write out the number (459 684 in this case) in a grid similar to that shown. Draw verticals down every *two* numbers to the left of where the decimal point would be. Ask what number *multiplied by itself* will be closest to but not greater than, 45 (in this case). This will be 6 times 6. Place a 6 above the 5 in the fifth column as shown. Now subtract the 36 from 45 and get the remainder of 9

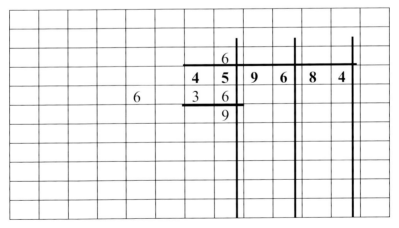

*Fig 2.23*

2. Now take the 6 in the very top row and *double* it. This makes 12. Place the 12 in the 3rd and 4th columns as shown. Now bring down the next two numbers, that is, 96 and place as seen in the sixth row. Now ask what number do we put above the six in

the 10th column which when added to 120 and then multiplied together will be less than 996. It will be 7, as $127 \times 7 = 889$. Place 7 above the 6 in the top row, and place 889 under 996 and subtract giving 107.

Fig 2.24

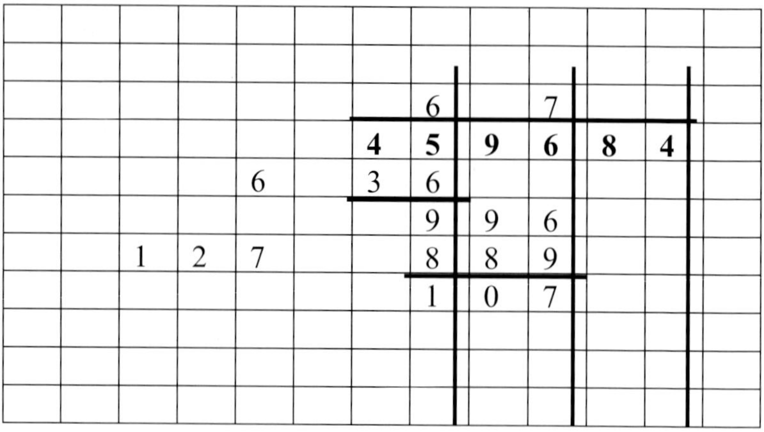

3. Repeat this process. Now take the 67 in the very top row and *double* it. That gives 134. Place the 134 in the second, third and forth columns as shown. Now bring down the next two numbers, that is, 84 and place as seen in the eighth row. Now ask what number do we put above the four in the twelfth column which when added to 1340 and then multiplied together will be less than or equal to 10784. It will be 8, as $1348 \times 8 = 10784$. Place 10784 under 10784 and subtract giving exactly **zero**.

Fig 2.25

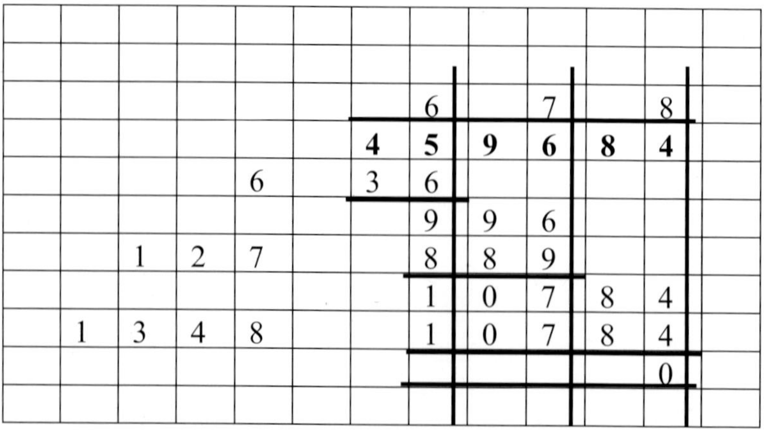

Since there is no remainder we can assert that the square root of 459684 is exactly 678. *Why* this method works is another matter.

But suppose we do not have a nice whole number with an exact square root? This method can neverless discover it. Ask the class to find the square root of 2 to *ten* decimal places. Most school calculators will only go to nine decimal places so to demand ten poses a wee problem! But no — we have the above algorithm. If I recall rightly I did do this once with a Year 7 group. They can be led through the process.

### Exercise 32 *The square root of any number (in particular √2)*

1. Find the square root of 2 to ten decimal places. Ouch! We start exactly as above, with placing the 2 in front of a decimal point but this time putting 22 (yes) zeros after the decimal point.

*Fig 2.26*

2. Now we ask what number times itself will go into 2, answer $1 \times 1 = 1$. So place the 1 above the 2 and place a 1 in line with the row below the 2 and also place the multiple of $1 \times 1 = 1$ immediately under the 2 as well.

*Fig 2.27*

3. Now subtract the 1 from the 2. This gives 1 again. Now bring down two zeros making 100. Place *double* the very top 1 on the left-hand side and multiply by 10 making 20.

   Now test for a number less than 100 from $22 \times 2 = 44$, $23 \times 3 = 69$, and $24 \times 4 = 96$. Only 24 and 4 gives a number with a remainder less than 24 when subtracted from 100. This is a good exercise in *estimating*.

*Fig 2.28*

4. Subtract 96 from 100, this leaves a remainder of 4. Now again bring down two zeros making 400. Continue this whole process.

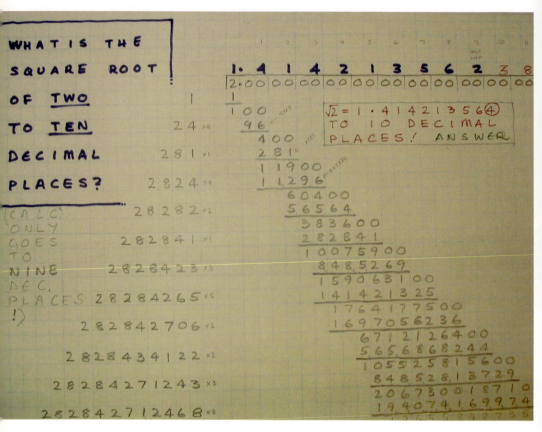

Fig 2.29 Calculation for the square root of two

The result up to 11 decimal places is shown here (why eleven and not just ten?).

This is an excellent exercise in numerical accuracy and the processes involved are addition, subtraction, multiplication and estimating. It is easy to make a mistake, although a check with the calculator can keep us on the right track until the last minute, or rather, last two decimal places required.

*Exercise 33  Square roots (but no calculator please)*

1. Find the square root of 1 522 756 exactly.    1234

2. Find the square root of 2.618 to two decimal places.    1.62

3. Find the square root of three to three decimal places.    1.732

4. Find the square root of 3 to *ten* decimal places (over the weekend!)
   1.732 050 807 568 9...... which rounds off to 1.732 050 807 6 to ten places

5. Can you find the square root of –1 (is this even *possible?*)
   No

## Pythagoras theorem

This is usually stated as:

> For a right-angle triangle, the square on the hypotenuse is equal to the sum of the squares on the other two sides.

As can be seen there are many right-angle triangles for which we do not need fancy ways, or even a calculator, to find a square root. This is when we already know what the square of a number is.

Some examples of such familiar triads are:

3 – 4 – 5,  5 – 12 – 13,  7 – 24 – 25,

and multiples of these such as 6 – 8 – 10,  21 – 72 – 75.

We also saw how we can find a good approximation to the square root of a number if there is no whole number solution.

This is all well and good. But mathematician types need to know things for certain, for all cases and forever! Which is why they make such a fuss about *proofs*. To show a few special cases can be called *demonstration* but it is not proof. So we appeal to the exigencies of logic. We trust our thinking.

Algebraic proofs belong to later years so only some of those that involve constructions will be shown here.

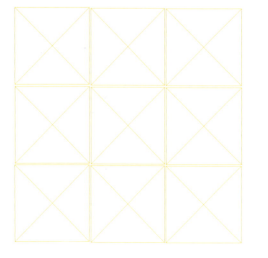

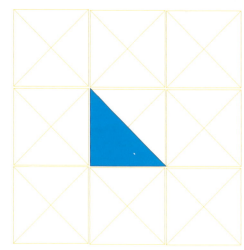

*For a right-angle triangle ...*

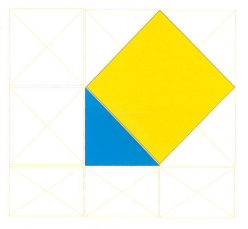

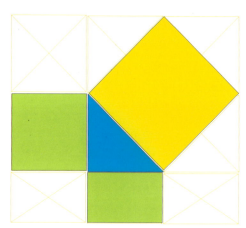

*... the square on the hypotenuse ...*

*... is equal to the sum of the squares on the other two sides*

*Fig 2.30 A floor tile demonstration*

## Demonstration

Consider the floor tiles ... one nice demonstration is that visible when viewing *nine* floor tiles where each square floor tile is decorated with *four* isosceles triangles.

# PYTHAGORAS THEOREM.

A theorem, usually attributed to Pythagoras
but known to much of the Ancients worlds
— was also called — the Forty Seventh
Proposition (of Euclid) — the Theorem of the
Bride — Dulcarnan (two horned) — Carpenters
Theorem — Windmill Theorem — the Franciscan
Cowl. Many postage stamps honor it!

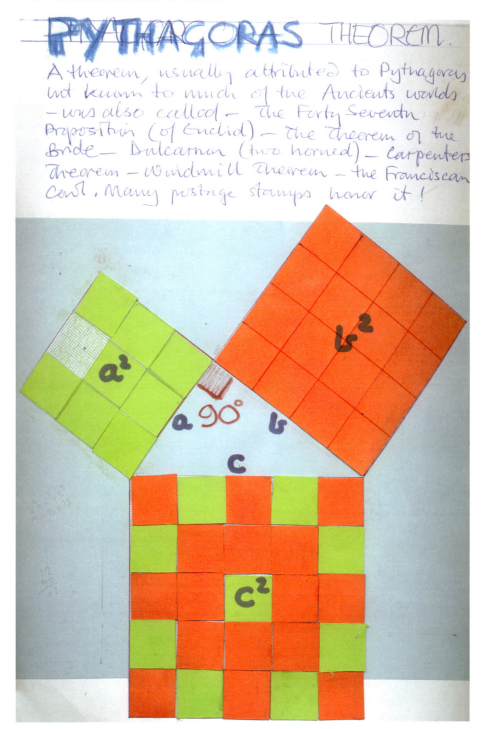

*Fig 2.31  345 Right-angle triangle*

*Exercise 34   Pythagoras 3 – 4 – 5 triangle demonstrated*

1. Draw a 3 cm × 3 cm square on yellow card.

    That is: $3^2 = 3 \times 3 = 9$

    This can be represented by a square 3 units by 3 units.

*Fig 2.32*

2. Draw a 4 cm × 4 cm square on blue card

    That is:  $4^2 = 4 \times 4 = 16$

    This can be represented by a square 4 units by 4 units.

*Fig 2.33*

3. Cut the squares into nine and sixteen unit squares respectively.

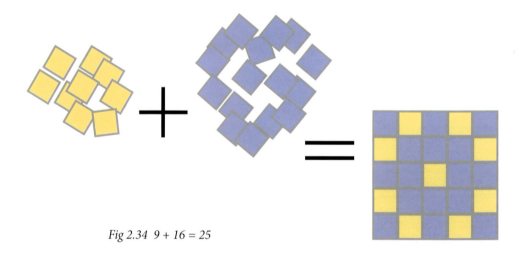

*Fig 2.34  9 + 16 = 25*

4. Let these two squares be disassembled and then the 9 plus 16 unit squares be reassembled into a single rectangle. It is found that one of these rectangles is a *square* with each side five units as in Fig 2.34.

Put these three squares together as in Fig 2.31 and it transpires that the angle opposite the *longest* side is a right-angle. We literally see that *for a right-angle triangle the sum of the squares of the sides is equal to the square on the hypotenuse.*

Hypotenuse functionally means the longest side and literally 'stretching under' from the Greek. The other two sides are called the 'legs,' or simply 'sides.'

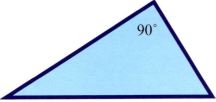

*Fig 2.35 'stretching under,' hypotenuse*

There is a nice website out there (*www.cut-the-knot.com/pythagoras/*) by Alexander Bogomolny where fifty-four proofs of this theorem are described. The writer also mentions a book by Elisha Scott Loomis, an early twentieth-century professor, with 367 proofs of Pythagoras theorem!

## *Bhaskara's proof*

Most theorems have an algebraic component, so may be difficult for Class 7. One that is neat is that attributed to Bhaskara in about the year 1150. See Fig 2.37.

*Exercise 35   Bhaskara's proof of Pythagoras Theorem*

1. Take four right-angle triangles triangles all exactly the same size.

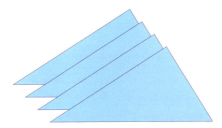

*Fig 2.36*

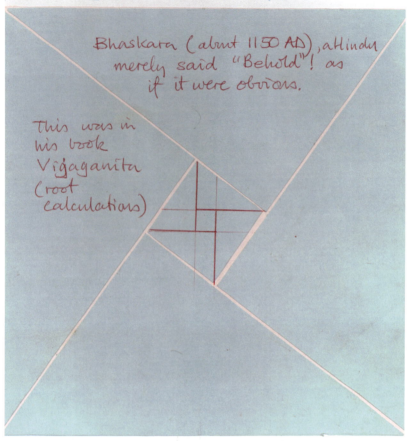

Bhaskara (about 1150 AD), a Hindu merely said "Behold"! as if it were obvious.

This was in his book Vijaganita (root calculations)

Fig 2.37 ... and Bhaskara said 'Behold'!

2. Now rearrange these to form a *square* where the side of the square is composed of the two different shorter legs of the initial triangles. This leaves a hole in the middle.

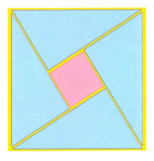

*Fig 2.38*

3. Label the sides of a triangle *a*, *b* and *c*.

4. Give the small square length *d*.

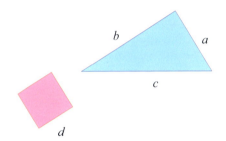

*Fig 2.39*          *d*

5. We note that by summing the triangles and the small square:

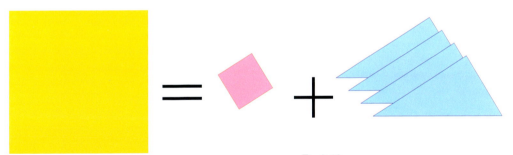

*Fig 2.40*

6. And that $d = b - a$ (look closely at 2 above).
 So that $d^2 = (b - a)^2$

 Also we note that the area of *one* blue triangle is $\dfrac{a \times b}{2}$

Putting the above in symbolic form we can say that:

$$c \times c = d \times d + 4 \frac{(a \times b)}{2}$$

or     $c^2 = d^2 \quad + \quad 2ab$

thus   $c^2 = (b - a)^2 + 2ab$

and multiplying out we get:

$c^2 = b(b - a) \; - a(b - a) \; + 2ab$

$c^2 = b^2 - ab \; - ab + a^2 + 2ab$

$c^2 = b^2 - 2ab + a^2 + 2ab$

$c^2 = b^2 + a^2$

QED — That is: *quod erat demonstrandum* (which was to be demonstrated) or quite enough done!

Hence:       $$a^2 + b^2 = c^2$$

as the formula is usually written.

Finally some students work over the years

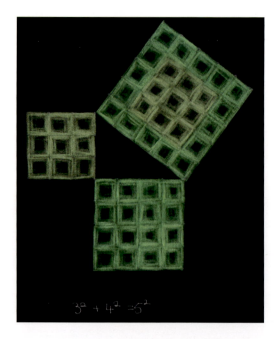

*Fig 2.41*

# PYTHAGORUS THEOREM.

A Right Angle Scalene Triangle

*Fig 2.42*

Fig 2.43

One theory is that Pythagoras discovered his famous theorem on his way to the bath...

...He saw the solution in the floor tiles of the bath house.

*Fig 2.44*

# 3. Platonic Solids

Class 8 (13–14 year olds) is the time for the student when there is a movement from the planar to the spatial. They have hopefully been modelling forms in clay, mud, plasticene or beeswax for some years, and making models of all sorts. Now this can be given an abstract turn — in the best sense — where the forms are seen to have an inherent truth of their own, irrespective of their physical manifestation.

They can now be rendered susceptible to calculation and an expression of accuracy and mutual relationships which are not immediately visible in the work and play with such forms as may have been the case until now. That accurate calculations and careful drawing are needed to bring about satisfactory forms now becomes evident. The students are capable of this too now. Their modelling and drawing work are often exceptional.

*Fig 3.1  Dodecahedron. Glass model by Christel Post.*

PLATONIC

SOLIDS

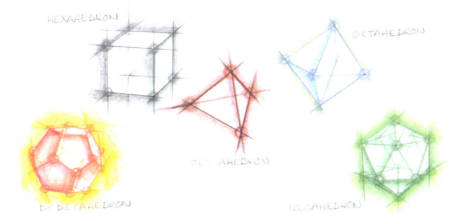

*Fig 3.3 The five Platonic solids*

Here, in this main lesson they approach *form* — and it is the most simple and regular forms possible which are transformable and interrelated, also in space. They can even be seen in the world of nature — although we have to *look*. Nature does not give up her secrets too readily, even with the most basic structures. *We* have to do some work to see the 'revealed secret.'

## A title page

For a cover I have used the famous idea expressed by Johannes Kepler with his *Cosmographicum*. This is the notion that the very planets are arranged within vast spheres, all of which are centred on the sun and that the spacing is determined by all five Platonic solids. For instance, between the planetary spheres of Saturn and Jupiter is the cube or *hexahedron,* between Jupiter and Mars is the *tetrahedron* and so on. His outermost planet was Saturn because the further planets had not been discovered yet. To build such a model, or at least some of it, would be quite a challenge for a class of this age, with the forms of greatly different side lengths for each solid.

But what are these Platonic solids and why do they appear significant at this stage of the student's career? Both questions will be considered in due course. So we know what we are talking about, Fig 3.3 shows the five Platonic solids.

◄ *Fig 3.2 A title page*

*Fig 3.4  Icosahedral die from ancient Egypt in the British Museum.*
*(free sketch by JB of die from Critchlow,* Time Stands Still, 145)

## Platonic solids historically

Cultures of the past have in various ways and for various reasons given various expressions to how they have regarded these five solids. An *icosahedral* die has been found in Egypt (Critchlow, *Time Stands Still*, 145).

In north-east Scotland many small carved-in-rock sphere-like figures have been found. The manner of the carving on many of them seems to indicate a clear awareness of the five solids, although they are not limited to these. Upwards of 500 have been found. What they were *for* is unknown (we always want a reason, do we not?). Who did them is unknown. When they were carved is unknown, although the archeologists estimate about 2500 BC, that is 4 500 years ago.

Now what were our so-called primitive Neanderthals doing in making numerous Platonic solids well before Plato even lived? Nor are they presented in the way we are familiar. Two examples of the cube (or *hexahedron*) are shown here (Fig 3.5a and b). A

*Fig 3.6  An introductory page* ➤

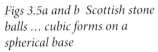

Figs 3.5a and b  Scottish stone
balls ... cubic forms on a
spherical base

cubic form is obviously intended and suggested by appropriately
sized and oriented rings on a basic sphere.

Plato described them in his *Timaeus* and gave to each form
a connection to the five elements of his time: earth, water,

# Platonic Solids : Introduktion

Mother nature demonstrates the greatest variety
of crystalline, facetted forms. Many tiny
plant animals and virus forms have been found
to be regular geometric constructions. If we
wish to study these we need to discover some
of the laws of Space and Geometry of Solids.

We can start with the simplest possible regular
shapes in space. There are five of these and
they have been named the PLATONIC SOLIDS
since Plato was one of the first to describe them.

Early Egypt was aware of them — there have been
found ICOSAHEDRAL dice — and they were
later fully described in the 13th Book of Euclid.
Even in Neolithic Scotland, small granite forms
have been discovered with represent these five
solids within the sphere. Their use is a mystery.

All of the forms (except one) can be made
from TESSELLATIONS of the plane. One of
the forms is the CUBE or HEXAHEDRON
and this is constructed from a net of squares
drawn from a tiled or tessellated plane
of squares. We introduce the CUBE with
a model made by folding in a special
way, a net of squares.

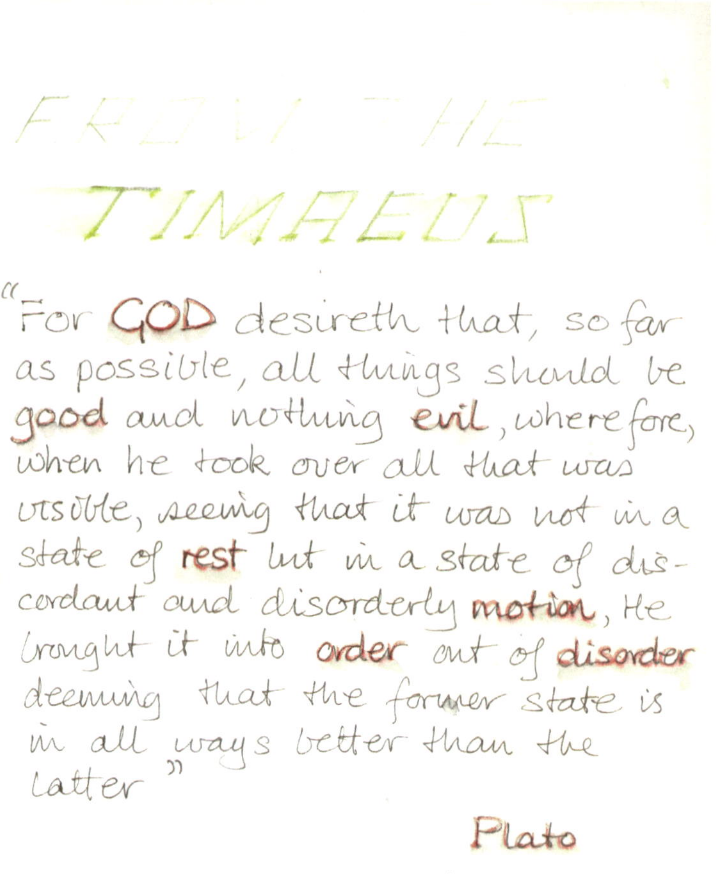

## FROM THE
## TIMAEUS

"For GOD desireth that, so far as possible, all things should be good and nothing evil, wherefore, when he took over all that was visible, seeing that it was not in a state of rest but in a state of discordant and disorderly motion, He brought it into order out of disorder deeming that the former state is in all ways better than the latter"

Plato

*Fig 3.7  Plato — on what 'God desireth'*

air, fire and the universal ether. Did he mean space with this universal? For if he did then he may have been extraordinarily prescient as in a recent article in *Nature* there is a hint that space itself may be in some sense dodecahedral (*Nature*, 2003 Oct 9, 425, 593–95).

This notion arises from calculations of how the background radiation, which many scientists believe expanded from earlier times, does not have the broadest waves it is supposed to have for a flat universe. So — if not flat — then it must be shaped! The writers propose a Poincaré *dodecahedral* space. So is space dodecahedral in some sense after all?

## *Planar figures*

The calculation and accurate construction of the first three *planar* figures with pencil, compass and ruler in the plane is a relatively easy first task. First we note the division of the circle by the counting numbers, three, four and five. Why these three only becomes apparent when we wish to make regular polygons.

Dividing the circle by 'one,' 360° / 1 = 360°, that is so-called 'trivial,' as we can't have a *one* sided figure. And 360° / 2 = 180°, and that is incapable of creating an enclosed form — it would have to have only *two* straight sides, hence impossible also. But 360° / 3 = 120°, and if these angles are at the *centre* of the circle

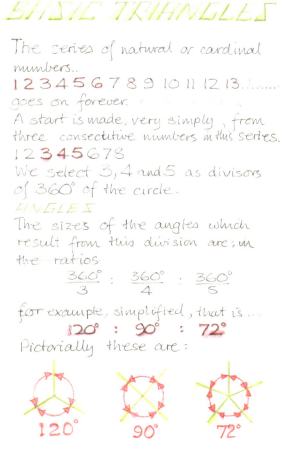

Fig 3.8 Divisions

and rays go from this centre to meet a surrounding circle then if we join these meeting pints we have the *equilateral* triangle. Doing a similar thing with four gives $360°/4 = 90°$ and leads to a *square*. And with five gives $360°/5 = 72°$ which readily leads to a regular *pentagon*.

## Three special right-angle triangles

If this is analysed a bit further then we can see three particular right-angle triangles emerging. These crop up in many contexts in later work. The three are shown in Fig 3.9 with construction methods. The third triangle should be definitely practiced before presentation to a class.

*Exercise 1   Draw the three right-angle triangles*

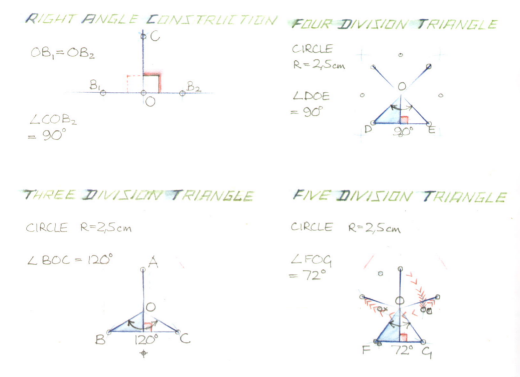

Fig 3.9 *Constructing a right-angle, and the three right-angle triangles with angles 120°, 90° and 72° at circle center*

*Exercise 2   Three polygons*

Building the three polygons, that is the equilateral triangle, square and pentagon together in the circle (see Fig 3.10).

   We can also show the three triangles here too.

EQUILATERAL TRIANGLE AND SQUARE AND PENTAGON

Equilateral triangle, all sides equal. Square, a regular quadrilateral. Regular pentagon

a five sided figure with all sides equal.
   The 3, 4 & 5 sided regular shapes shown in the circle are the three forms required to make the five solids first described by Plato. He built the three figures from right angled triangles.

*Fig 3.10   Polygons in the circle*

# CUBE-BY FOLDING

The net below
was constructed,
cut out and

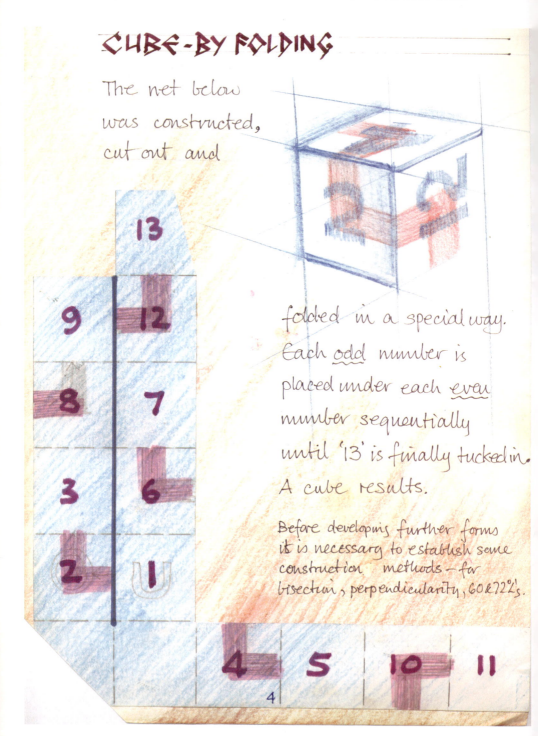

folded in a special way.
Each _odd_ number is
placed under each _even_
number sequentially
until '13' is finally tucked in.
A cube results.

Before developing further forms
it is necessary to establish some
construction methods — for
bisection, perpendicularity, 60 & 72°'s.

Fig 3.11  Cube by folding

## Cube by folding

So far all this has been in the plane. An interesting first modelling exercise can be the cube — by *folding*. This method is from Cundy and Rollett, *Mathematical Models,* an excellent reference book for this sort of work which includes many other folding constructions.

*Exercise 3   Creating a cube by simply folding over and over again*

It is good to make this model and the students can contrive all sorts of ways to present it and it can provide a good exercise in accuracy as all the fifteen squares should *be* square and also the *same* size.

One could get the students to make the cubes all different sizes but a good cooperative exercise would be to agree on a side length (say 5 cm) and make them all the same. Then they could be used as 'bricks' and see how *even* a wall (or some other construct) could be built up from the whole class's work. It is salutary and satisfying if all have to make a contribution to the construction.

## Some details of the three triangles

In Fig 3.12 the *angles* of the three triangles are discovered. This is simple to do as we know that *the sum of the angles of any planar triangle is always 180°.*

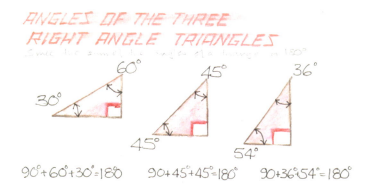

Fig 3.12  Angles of the three triangles

*Exercise 4   Finding the unknown angle of a triangle*

Find the unknown angles in the triangles:

a) When the triangle has angles of  90° and 24°. What is the third
   angle?    66°
b) Given that one angle is 35.5° and another is 64.5° find the final
   one.    80°
c) If two angles are both 63° what is the third?    54°
d) In the obtuse angle triangle *ABC* there are angles of 12° 45' and
   3° 56'. What is the largest angle? Sketch the triangle approximately to scale.    163° 19'
e) On a base *AB*, angle *CAB* = 100° and angle *DBA* = 110°. Can you
   form a triangle from this data?
   Yes, produce *CA* and produce *DA* until they meet at a point *E*.
   The triangle *ABE* is formed.

*Exercise 5   Pythagoras theorem*

In *Mathematics Around Us* the Theorem of Pythagoras was
explored. Simply stated in symbols it is:

$$a^2 + b^2 = c^2$$

where *c* is the longest or *hypotenuse* side which is always opposite the right-angle and *a* and *b* are the other two sides. We will
need this relationship in working out sides, angles, areas and
volumes of the Platonic and other solids.

   Using this relationship, find the unknown side lengths in the
*right-angle* triangles stated below (it may help to sketch the triangles).

a) When the triangle has one side 3 units, another as 4 units and the
   right-angle is *between* these two sides (or the included angle),
   find the hypotenuse.
   5 units

b) If the longest side (hypotenuse) of a right-angle triangle is 2 cm
   and the shortest is 1 cm, find the third side length to three decimal

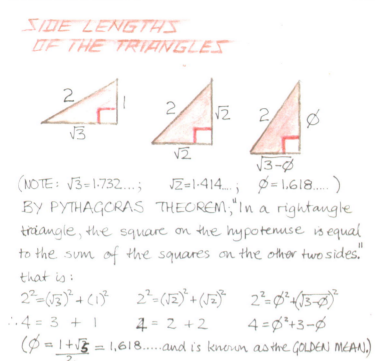

(NOTE: $\sqrt{3}=1.732...$;     $\sqrt{2}=1.414...$;     $\emptyset=1.618.....$ )

BY PYTHAGORAS THEOREM; "In a rightangle triangle, the square on the hypotenuse is equal to the sum of the squares on the other two sides."

that is:

$$2^2=(\sqrt{3})^2+(1)^2 \qquad 2^2=(\sqrt{2})^2+(\sqrt{2})^2 \qquad 2^2=\emptyset^2+(\sqrt{3-\emptyset})^2$$

$$\therefore 4 = 3 + 1 \qquad\qquad 4 = 2 + 2 \qquad\qquad 4 = \emptyset^2+3-\emptyset$$

$$(\emptyset = \frac{1+\sqrt{5}}{2} = 1.618.....and \ is \ known \ as \ the \ GOLDEN \ MEAN.)$$

Fig 3.13  Side lengths of the triangles


places (see panel for help).

> 1.732 cm

c)  The included angle between two sides is 90°. One of the sides is 13.33 units and the other is 3 times this. What is the longest side (give answer to nearest whole number)

> 42 units

d) A bull ant is 4 mm high. Its shadow on the ground is 13.5 mm. How long (to the nearest mm) is it from the top of the bull ant to the tip of the shadow? (and who on earth would want to know?)

> 14 mm

e) Is the triangle with sides 5 km, 13 km and 12 km a right-angle triangle? If so, *show* it.

> $5^2 + 12^2 = 25 + 144 = 169 = 13^2$   QED

# BOWLS and SADDLES

## BOWL FORM

What occurs when we "vary the extension of the perimeter with respect to the centre of a plane form?".

If a flat disc of some plastic material is worked with on the **INSIDE** so that it is thinned, then gradually a **BOWL** form developes.

DISC ⟹ BOWL FORM

## SADDLE FORM

On the other hand, if a similar flat disc of material is worked with on the **OUTSIDE** then gradually a buckling takes place, the edges crinkle causing a **SADDLE** form to develop.

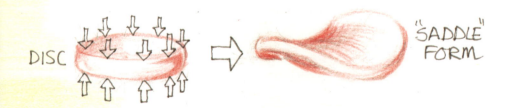

DISC ⟹ "SADDLE" FORM

## Bowls and saddles

In this section quite another aspect of planar forms is reviewed. It is based on some work by Cedric Lethbridge some twenty years ago which I have since developed. Later we come back to the various simple planar forms. But how do these planar or surface forms work together? This may seem an odd question but it leads quite a way.

If we imagine a homogeneous material formed into a circular disc of reasonable thickness we can ask what would happen if it is worked with *in the centre,* and if it were worked with on the periphery.

If the circle of plastic material is manipulated so as to *thin* it in the middle a *bowl* form will result. See Fig 3.14.

On the other hand, if the material is manipulated at the *edge, saddle* forms emerge (see the bottom of Fig 3.14). It is not too hard to see that, if the material is *not* plastic, certain constraints apply and only particular form possibilities exist. Imagine we can only deal with stiff material that can only flex at hinges or at straight lines.

Cedric Lethbridge noticed that this has consequences for the Platonic solids. If a specific number of regular polygons (triangles, squares, pentagons) are brought together then particular cases of the Platonic solids are formed.

Imagine the regular hexagon, made as it is of six equilateral triangles. Remove *two,* and the resulting four can be folded up to form part of an *octahedron.* Remove *three* such triangles and the three remaining can become a three-sided pyramid and this then forms part of the *tetrahedron.* Remove only one and the five triangles turn into one corner of an *icosahedron.*

Life gets interesting when triangles are *added* to the six. The arrangement can now no longer remain flat nor can it form convex (hollow) shapes such as the bowls.

◄ *Fig 3.14 Bowls and saddles*

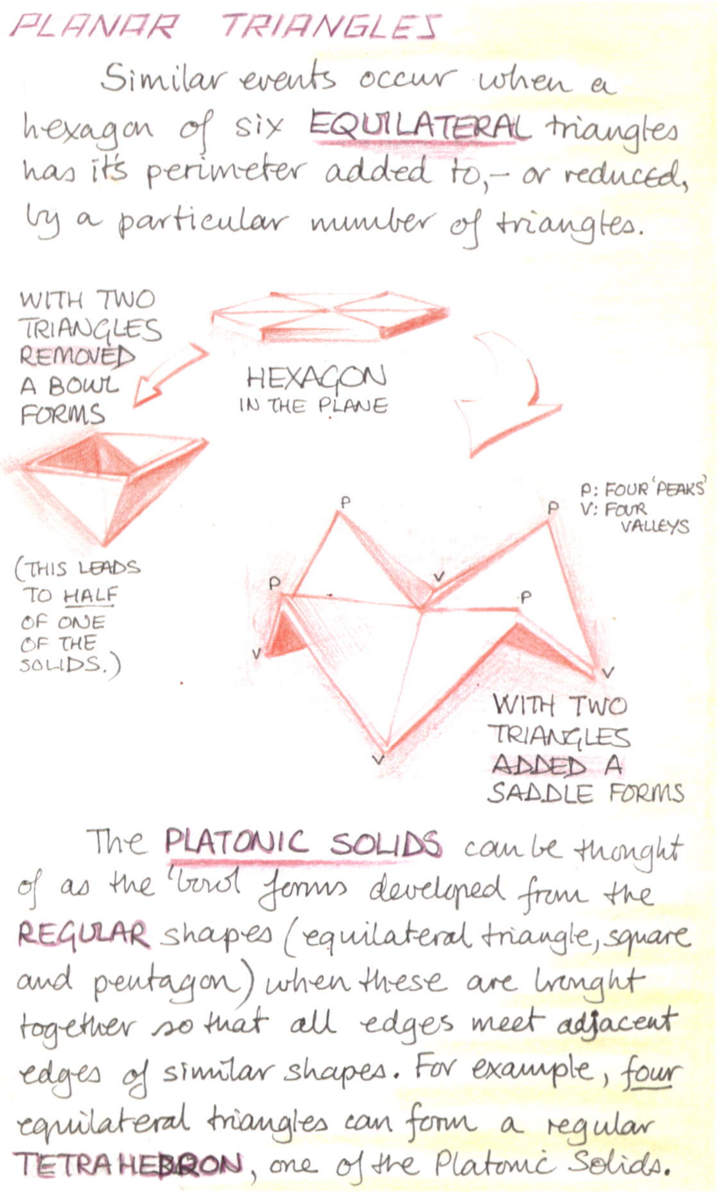

## PLANAR TRIANGLES

Similar events occur when a hexagon of six **EQUILATERAL** triangles has its perimeter added to, – or reduced, by a particular number of triangles.

WITH TWO TRIANGLES REMOVED A BOWL FORMS

HEXAGON IN THE PLANE

P: FOUR 'PEAKS'
V: FOUR VALLEYS

(THIS LEADS TO HALF OF ONE OF THE SOLIDS.)

WITH TWO TRIANGLES ADDED A SADDLE FORMS

The **PLATONIC SOLIDS** can be thought of as the 'bowl' forms developed from the **REGULAR** shapes (equilateral triangle, square and pentagon) when these are brought together so that all edges meet adjacent edges of similar shapes. For example, four equilateral triangles can form a regular **TETRAHEDRON**, one of the Platonic Solids.

*Fig 3.15 'Bowls' from regular shapes*

If regular radial symmetry is to be retained then *even* numbers of triangles need to be added. Fig 3.15 shows two triangles added and the buckled form that arises is illustrated. This buckling is a phenomena that appears in many places as soon as the attempt is made to insert more material into a planar of surface like structure.

Fig 3.16  Crinkling  in a local Australian Banksia leaf

## Leaves — and their holes and crinkles

In my own work on morphology (the study of form) it is evident that the leaves of plants are usually surface-like entities. They are often quite flat, but when more material tries to fit into this flattish leaf structure we get buckling. Again and again we get leaf edges with buckles or ripples or crinkling or hollowing or even holes. Some examples of crinkling, holes and internal buckling are shown in Figs 3.16–18.

Fig 3.17 (left)  Holes in a large Monsteria leaf

Fig 3.18 (right)  Internal buckling in a non-fruiting Passion Fruit leaf

In many vegetative forms we see a relatively consistent buckling or ripple. Less frequently I have observed part of the plane form unfilled — leaving gaps or holes.

With many mineral forms we see enveloping surfaces structured as enclosing planes, as edges which join the planar surfaces, as corners or vertices where three planes and edges meet. Here the folded forms meet neatly without buckling or holes. All the Platonic solids are like this.

## Point and periphery

With regular solids, such as the Platonic solids these enveloping lines, points and planes surround a central point, a *centre of gravity.*

Does this imply its dual, a *periphery of levity?* In later years we get to actually use this distant plane. It is already implied as an absolute plane with our regular forms, the Platonic solids, it is a geometric necessity. This is evident when we attempt to draw transformations of — for instance — the *cuboid* (a term I use for a geometrically consistent cube-like form structure, see Fig 3.19) and we have to use an *external* plane *ABC* to do the construction and an *internal* point arises of necessity.

We find that such mineral-like forms always relate to a geometric point *centre* and a geometric planar *periphery.*

*Fig 3.19 Cuboid form between plane ABC and inner point*

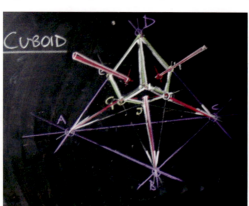

## The tetrahedron

Going back now to the most elementary of all forms in space, which is not the sphere (in this context) but rather the form which arises from the *minimum* number of points, lines and planes which can enclose — or exclude — a volume. This is the structure known as the *tetrahedron.*

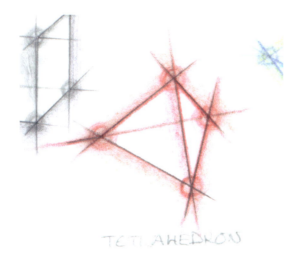

Fig 3.20 The tetrahedron

There could hardly be a more significant, yet simple, structure. Here there are four points, six lines and four planes. Any less and we do not have a form as such which can enclose any space. There are many mysteries to this form (see Lawrence Edwards) but here we take the most regular structure in its most easily constructed aspect.

This is where all the faces are equilateral triangles. From a planar net we can construct this figure. This gives a clean, neat model. But we can also emphasize the point-wise aspect (put four tennis or soccer balls together!) or the line-wise aspect as well. Models from sticks, straws, dowelling can be made easily. Really one aspect is no more important than the other. All are 'true,' inherent and integral.

*Exercise 6   Constructing a tetrahedron*

1. First draw a line horizontally across the page and mark two points
   on it (say) 5 cm apart — call them *A* and *B*.

Fig 3.21

2. Next, with compasses set at the same radius, draw two intersecting arcs from *A* and *B* to meet at *C*.

Fig 3.22

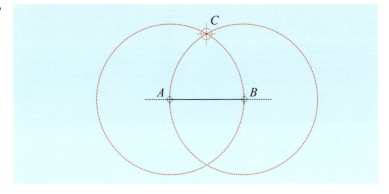

3. Now draw in the lines *AC* and *BC*. This creates the equilateral
   triangle *ABC*.

4. To form the net for the tetrahedron we put a further circle of
   radius 5 cm centred on *C*.

Fig 3.23

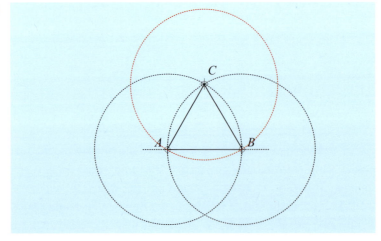

Fig 3.24  Tetrahedron net ➤

# TETRAHEDRON

The NET for the TETRAHEDRON can be drawn from the equilateral TILING of the plane. Four equilateral triangles are required. The

The tetra- hedron so form- -ed has four planes, six edges, and four vertices. This form is a four faced pyramid.

...that solid which has taken the form of a pyramid, shall be the element and seed of fire....." It is

Plato considered that this form could be related to fire. "Thus.....

the form which requires the min- imum number of vertices, edges

5. Draw in the *DCE, EBF* and *FAD* and we have the required net. All that needs to be added is the tabs for glueing the form together. Get the students to figure out where they are best placed. Do not make them too narrow (a common error). If the triangles are of 5 cm side then the tabs are best at least 1 cm wide.

*Fig 3.25*

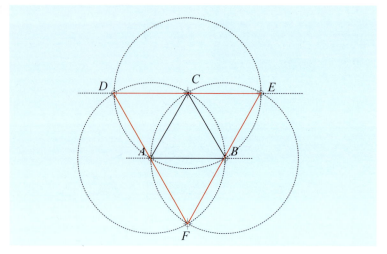

We see that the net can be further extended in all directions across the page. This is important to notice as this is needed again for two other Platonic solids. So a tiling or tessellation of the plane is created, Fig 3.24 (p.27).

Lines that need to be *scored* so that the paper or card folds easily can be done with the back of the blade of a pair of scis-

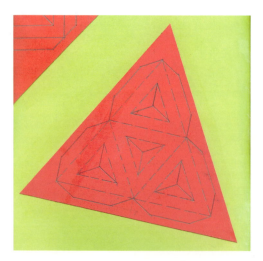

*Fig 3.26*

sors, a compass point held almost horizontally, a blunt modelling knife or even a fine biro point. Take care.

## Where are tetrahedra?

Where do we see this tetrahedron around us? It is in fact everywhere but still has to be looked for. The chemists and physicists tell us that in *carbon* there is a tetrahedral structure. It has around its centre four arms that reach out (known as bonds) and these always

Fig 3.27 (Continues from Fig 3.24)  *The form of fire according to Plato*

form some kind of tetrahedron. This can form a most rigid structure in space resulting in the hardest natural material, diamond.

Even the huge earth body we live on is slightly tetrahedral.

It would seem even at this level that the tetrahedron is a pretty fundamental form if we walk about on it every day, and it is the basis, in carbon, for every living cell in our bodies and in all forms of life. It is a significant part of 'greenhouse' gases, both carbon dioxide and methane.

Note the numbers:
    4 points
    6 edges
    4 planes

This is a clue already to an important theorem due to Euler — but more later when other solids have been examined.

*Fig 3.28  Fire from the ground at Olympos, Turkey (photo by Bahram Saba)*

As for the *form* of fire, a picture of ancient flames gives a wonderful expression of raw fire and flame issuing from the ground at Olympos, Turkey.

Were the Neolithic Scots in some way aware of this? Why make all these careful carvings on a stone ball, in a form very strongly suggestive of a tetrahedron cut into a sphere? Was there some awareness of the significance of the tetrahedral form in our world — or were these elaborate carvings merely used as doorstops? I think not.

These forms remain a mystery to the archeologists. Most of the Platonic solids are represented among the many finds made in Scotland (most near Aberdeen) according to Keith Critchlow.

*Fig 3.29  Carved tetrahedral Neolithic stone ball (from Critchlow,* Time Stands Still)

*Fig 3.30 Solids galore — student's work at an exhibition*

## Octahedrons

This form is also constructed from equilateral triangles. Instead of four planes, or surfaces, as in the last form, there are now eight. Also we have six points or vertices not four as with the tetrahedron.

Is there any system in the numbers and kinds of elements? There is, and Euler discovered it. He found a simple relationship between the points, lines and planes of any convex prismatic body. See if the students can discover it for themselves from the numbers of points, lines and planes. And where do these forms appear in nature?

## Octahedral net

Again octahedral forms can be constructed to give a point-wise emphasis or line-wise etc. Here a design for a planar model is offered. Since the form is comprised of all equilateral triangles we find that all we have to do is extend the use of the tiled plane used for the tetrahedron in a specific way.

This is easy to do but needs to be thought through as to exactly which triangles to use and where to put all the glue tabs and what lines to score. It is a little challenge to give the students the tiling (or tessellation) and get them to try to select a suitable set of eight. After that they need to draw the tiling for themselves from scratch. I have always tried to get the students to draw their own nets, even though the black-line master might appear an easy way out, as it is a good exercise in accuracy. In that way they are a part of the whole process, developing skills at the same time.

*Fig 3.31 Octahedron net*

*Exercise 7  Making a model of an octahedron*

1. Draw the tessellation to cover the page using circles of 5 cm radius (for instance) and select the set of triangles shown in Fig 3.31 (p.137). This is good practice for making the icosahedron which comes later as it requires even more of the tiled plane.

2. Draw in appropriate gluing tabs. This takes a little thought to not double up on the tabs.

3. Score along the dotted lines — including along tab lines (not shown in sketch).

4. Cut out the net (taking care not to cut off the tabs — easily done).

5. Fold along the score lines forming the octahedron itself.

6. Glue the tabs to appropriate neighbouring triangles. Sometimes the last tab can be tricky but one device is to hold the complex together with small bits of masking tape or equivalent while the glue dries. Then *carefully* remove it.

*Exercise 8  Making a model of an octahedron by folding*

This method is also from Cundy and Rollet. The net is shown in Fig 3.32. It is still made from an equilateral triangle tessellation but is somewhat more complicated.

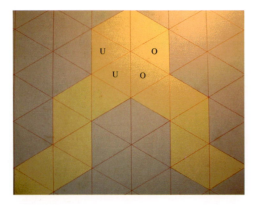

*Fig 3.32 Octahedral net for folding. O = over, U = under.*

It is a bit of a challenge to see how to fold the triangles after scoring and cutting out the yellow form, and also cut between *U* and *O* up to the *middle* of the hexagon.

No glue is needed as the last tab on the left tucks in nicely. Try it.

## More on octahedrons

Fluorite can be cleaved into octahedral shapes. Pyrites can be found with this crystal form.

The OCTAHEDRON was seen by Plato as the form representing the AIR. This solid is made from a net of eight regular triangles and is to be found in numerous crystal forms including fluorite and iron pyrites.

| | FACES | EDGES | VERTICES |
|---|---|---|---|
| OCTAHEDRON | 8 | 12 | 6 |

AIR

Fluorite Crystal

Spicules of a Sponge

RADIOLARIAN
C. OCTAHEDRUS

*Fig 3.33 Octahedron — an 'air' form in Plato's terminology*

*Fig 3.34 Cleaved fluorite*

Even in the organic realm there are very small radiolarian forms, presumably built of calcium, which Haeckel described in his famous *Art Forms in Nature* in about 1900.

For Plato this was the form he related to air. Why was this so? In *Timaeus* we find no easily understood explanation but a statement, 'So again with air: there is the brightest variety which we call aether, the muddiest which we call mist and darkness, ... which are produced by the unequal sizes of the triangles.' (*Timaeus*, 27)

With Fig 3.35 we see how the octahedron is able to be seen as a stacking of cubes (how many?)

This hints further at some connections of these two forms, since we can build the octahedron itself from a multiplicity of same size cubes.

And thus leads quite readily onto modelling the cube itself and exploring it as a form in its own right as well as possible transformations of the one form into the other.

*Fig 3.35 Cubes stacked to form an octahedron*

## The hexahedron (or cube)

The hexahedron is intimately related to the octahedron. Hence I have followed the octahedron model with the hexahedron model. Hexagon means six-sided and hexahedron means six-*faced.* It is of interest that all the Platonic solids, are named for their number of *faces,* not their points or their lines — which in their own way are no less important of course. Just a little less obvious. So the hexahedron has eight points, not six, so we do not generally give the same emphasis to corners as to faces. Not that it matters too much so long as we know what we are talking about. This naming is an artifact of convention after all. The points and lines *could* complain about unfair discrimination ... and it is somewhat symptomatic of our way of looking at things that these other aspects are given less importance.

How are these forms related to each other? They are *duals* of each other (more later), they can *transform* into each other and the numbers of elements are the same yet different. See the table below.

|                      | Faces | Edges | Vertices |
|----------------------|-------|-------|----------|
| Octahedron           | 8     | 12    | 6        |
| Hexahedron (Cube)    | 6     | 12    | 8        |

It seems there is some reciprocity in the two forms relationships. Geometrically this, too, becomes obvious but it is best to make the model first.

## Hexahedron net

There are a number of connected layouts that will give a net from which the hexahedron can be made. Only one is shown here. All are based on the *square.* The plane needs to be tiled with squares.

*Exercise 9    Hexahedron or cube*

Fig 3.36  *A net for the hexahedron or cube*

Draw the net of the hexahedron as shown in Fig 3.36 and make a cube. There are other nets. One is in the form of a cross.

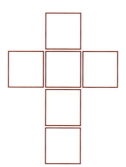

*Fig. 3.37*

1. First draw a line horizontally across the page and mark two points, *A* and *B*, on it 6 cm apart (for instance).

*Fig 3.38*

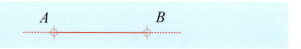

2. From now on I refer to Fig 3.39 for this exercise. Now draw two arcs from A and *B* to meet at *C* and *D*. This is one of the constructions for a right-angle. Join *CD* and this will then be a vertical line at *right-angles* to horizontal line *AB*.

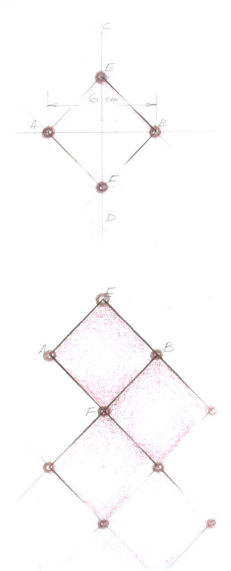

*Fig 3.39*

3. Then with compass set at *half* of *AB* describe a circle *AEBF* with centre where *AB* and *CD* cross.

4. Join the points *AE, EB, BF* and *FA*. This brings about a single square.

5. Produce (that is, extend) the lines *EB* and *AF* downwards to the right and also line *BF* down to the left.

6. With length *FB* as a radius mark a new point on *FB* produced as shown. Similarly do the same from *AF* and *EB*. It should now be possible to see how to continue the net down the page until at least six squares are drawn.

   This method gets less and less accurate with each extension as error accumulates. See if there is a more accurate method. (e.g. one could start with a much larger square and continually *sub*divide it.)

7. Draw in glue tabs, at least 1 cm wide.

8. Score along lines for all folds.

9. Cut the net out.

10. Fold and glue.

## *On hexahedrons ...*

The cube and rectangular prisms are with us in nature. Iron pyrites, salt and galena (a lead ore) can often approximate to the cubic structure, and are certainly built with mutually perpendicular faces.

Sometimes these seem to be so precise that people find it hard to believe that they are actual natural forms — believing them to be carefully machined and engineered. I have a wonderful sample of pyrites (Fig 3.40) which has been misinterpreted as something made by human artifice.

If the two pictures in Fig 3.40 are held *at least at arms length* and the eye makes the two red spots converge then the crystal may jump into apparent three dimensions ... it is worth a try, this one is quite effective.

Not only is the cube form the *dual* of the octahedron but *five* of them can fit into a dodecahedron. Further, *two* regular tetrahedrons can fit neatly into a cube.

One of the tetrahedrons is shown in the cube in the accompanying sketch in Fig 3.41.

▲ *Fig 3.40  Pyrites*

"To earth let us give the cube form for of the four kinds earth is the most immobile ..." so says Timaeus. The cube form, or hexahedron, is the only one of the Platonic solids which will fill space with no voids remaining. All its angles are right-angles and all its faces are square. The dual form to the hexahedron is the octahedron since for each face of former is a round point the other. There are crystals based in a cubic structure, for example fluorite, rock salt, galena pyrite, and the metallic copper gold and silver. The cube will fit five times into the dodecahedron and two interpenetrating tetrahedrons can be placed in it.

HEXAHEDRON
WITH TETRAHEDRON

| HEXAHEDRON | FACES | EDGES | VERTICES |
|---|---|---|---|
| | 6 | 12 | 8 |

*Fig 3.41*
*The hexahedron*
*and Plato etc* ➤

*Fig 3.42  Glass cubic model by Harriet S*

The cube form was an exercise for a student of glass-work. This was a practice piece from one of our students many years ago. It is an excellent rendering of the pure cube.

## Interpenetrating cube and octahedron

These two forms neatly interpenetrate each other. This sketch (Fig 3.43) is drawn in isometric projection. In this case the three infinitely distant perspective viewpoints are spaced around a central point at 120 degrees and the scale of all distances is the same in the three directions. Also called an isometric view.

Here is yet another relationship between these two forms. It makes for an interesting model as well.

*Fig 3.43  Interpenetrating forms of cube and octahedron*

*Fig 3.44  Dodecahedron*

*Fig 3.45  Icosahedron*

## Icosahedron and dodecahedron

These two forms are also related intimately to each other. In Fig 3.44 and Fig 3.45 we see two 'skeletal' renderings of the two forms.

Count up the faces (is it twelve?).

Count up the corners. How many are there?

And then determine the number of edges.

This is by far easier with a model in the hand rather than from a drawing!

Try counting the number of faces here. Does it echo any of the numbers for the dodecahedron?

Count the corners or vertices.

Finally check the number of edges or line segments.

Compare now, for the two forms, the numbers of edges, faces and vertices. Do we see any reciprocity?

|              | Faces | Edges | Vertices |
|--------------|-------|-------|----------|
| Icosahedron  | 20    | 30    | 12       |
| Dodecahedron | 12    | 30    | 20       |

If we have not figured out *Euler's Law* at this point, have another go. Hint: it goes something like this: Something + Something Else = Another Thing + 2. But what are the 'things'?

Faces (planes) + Vertices (points) = Edges (lines) + 2

There are secrets hidden in the very fabric of space, even for the simplest of forms. This is worth thinking about. There is nothing physical here in the concepts of these forms. We can confirm such concepts with physical models or even drawings but the *ideas* do not depend on them.

*Fig 3.46 Gaming dice — dungeons and dragons perhaps*

We may have individually discovered these notions by drawing in the first place but that does not mean the ideas and concepts are antecedent. *They* have been around for a long time and certainly long enough for Plato to think them and give significance to them. Perhaps even the neolithic inhabitants of northern Scotland knew of them as well as the ancient Egyptians with their dodecahedral dice. And we *still* use them for gaming!

## Icosahedron net

To make this model in planar surfaces the tiled surface of equiangular triangles is extended still further as shown in Fig 3.47.

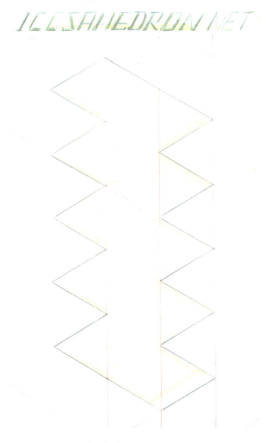

*Fig 3.47 Icosahedral net*

*Exercise 10   Icosahedron*

These instructions apply to Fig 3.48

1. Draw the net of equilateral triangles.

2. Decorate creatively!

3. Draw the tabs for gluing (yellow).
4. Score dotted lines (red).

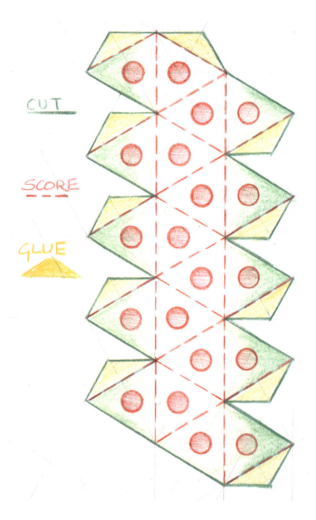

*Fig 3.48 Net for icosahedron, complete with, some decoration, glue tabs, dotted scoring lines and bold outline cutting profile*

5. Cut out bold profile (dark green).

6. Fold form to icosahedron shape.

7. Glue appropriate tabs, only one or two at a time, and hold together with small bits of masking tape lightly applied while glue is drying. Carefully remove masking tape.

8. Done and display.

This is the form in which the largest number of equilateral triangles can make a convex (enclosed) form. It has five triangles adjacent to each other. It is not hard to see that this is the most. For if there were *six* then all the six triangles are in a plane.

The ancient Scots even made models that could be interpreted as this form too.

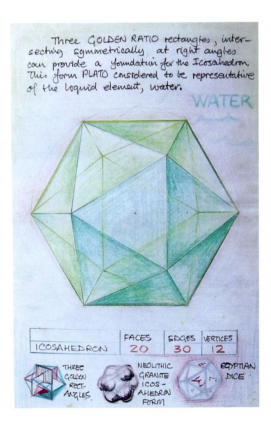

*Fig 3.49 Icosahedron*

## *The Golden section architecture of the icosahedron*

The icosahedron has a core structure and this is also significant. If three rectangles in three planes, mutually at right-angles, are slid together then their vertices form the corners of the equilateral triangles of an icosahedron.

That is, if the rectangles have sides in the Golden ratio, (golden section or proportion). The sides need to be in the ratios:

1 : 1.618 ... approximately, or $1 : (1 + \sqrt{5})/2$ exactly.

*Exercise 11    To construct a Golden rectangle*

The easiest construction for a Golden rectangle that I know is shown here.

1. First draw a square (red).

2. Divide the left side of the square into two equal lengths.

3. Using as radius the distance from this centre to an opposite corner strike an arc to the left side produced upwards.

*Fig 3.50  Three plywood Golden rectangles*

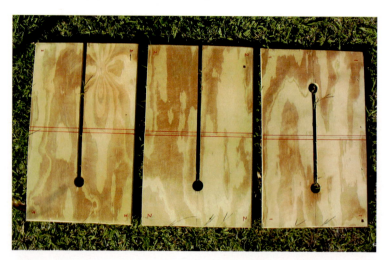

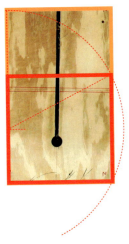

*Fig 3.51*

4. Where this arc strikes the left-hand produced line, the longer side of the golden rectangle is found.
5. Complete the golden rectangle.

6. To find the end point of the slit, find the centre point of the rectangle (diagonals) and mark the distance of one half of the shorter slit above and below.

7. Cut slits as shown.

*Fig 3.52 Icosahedron skeleton of three interpenetrating rectangles*

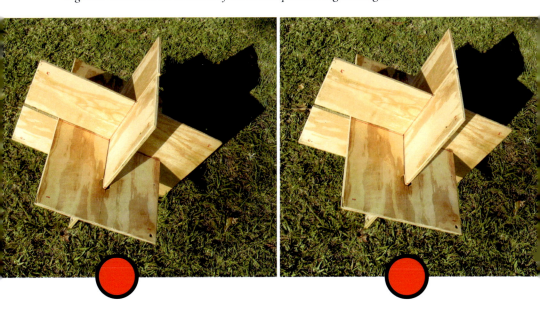

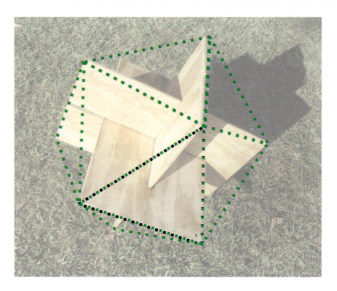

*Fig 3.53 Icosahedron skeleton*

The envelopment of the skeleton is shown in Fig 3.53 with dotted lines indicating where the icosahedron can be seen.

The proportions can be proved using Pythagoras Theorem. If the side length of the square is one unit length then the other side comes to approximately 1.618003 ... a somewhat difficult but good extension task perhaps.

Fig 3.52 shows how these three rectangular plywood bits fit together.

## Dodecahedron

The companion form to the icosahedron is the dodecahedron. This is the last and, in a way, most interesting of all the Platonic solids.

Plato had it that this structure is characterized by saying: 'There still remained a fifth construction, which god used for embroidering the constellations on the whole heaven' (*Timaeus*, 22).

It is also the *polar* form to the icosahedron. What this means is pictured in Fig 3.54. It is a good discernment exercise to see the one form or the other. Think 'green' and you see the icosahedron. Think 'mauve' and the dodecahedron can stand out for the

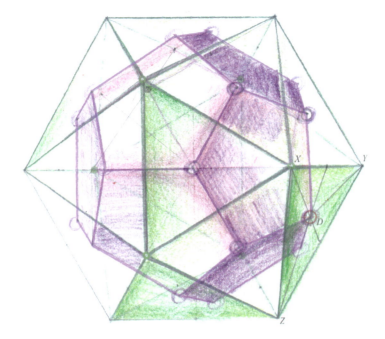

*Fig 3.54 Dual forms*

attention. We note that if the centres of all the adjacent triangles are joined with line segments then a range of *pentagonal* faces appears. The sum of these is the dodecahedron.

Needless to say if all the centres of this new form are taken and lines join these centres then we get back to the icosahedron. It is alternating icosahedrons and dodecahedrons infinitely all the way down to the unreachable centre point. There are two structures which I once read were considered the most frequent to be found in nature. These were the *helix* and the *dodecahedron*. This was stated by a writer in *New Scientist* many years ago. Amazingly some very small viruses also have these forms as their basic structure.

But there's more. Because we can go the other way too — and find bigger and bigger alternating forms — all the way out to the farthest reaches of the cosmos. Imagine my surprise recently when I read some correspondence in *Nature* magazine (referred to earlier) which suggested that, following an analysis of the background radiation data gleaned from satellite measurements, there appeared to be an interpretation that space was indeed *dodecahedral*. Was Plato right, in some sense, all along? Perhaps, but I have seen no follow-up on these ideas yet.

## *The Golden rectangle again ....*

And we meet those three Golden rectangles again. The centre of each pentagonal face touches one corner of the rectangles. So each rectangle has four pentagonal faces touching it. And 4 times 3 — as there are 3 rectangles — makes twelve. And our form has 12 faces does it not?

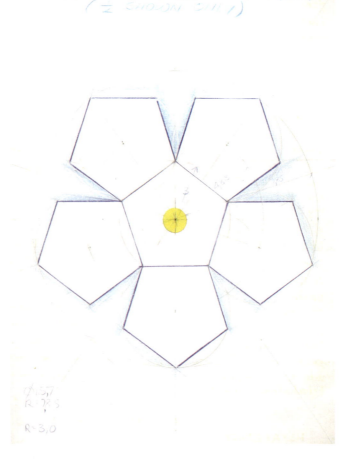

*Fig 3.55 Dodecahedron net — only half is shown*

## The dodecahedral net

This time we cannot tile the plane so easily. In fact it *cannot* be done with regular pentagons at all without leaving gaps or having any overlaps.

The net I choose to use is a single regular pentagon surrounded by five other identical ones as shown in Fig 3.55.

### Exercise 12   Model of the dodecahedron

The procedure is very similar to that for the icosahedron so it will not be repeated here.

Some care needs to be taken with glue tab design.

Also when forming the final figure it is helpful to make one complete 'bowl' of six connected pentagons first. Let glue dry properly. Then 'weave' the second six pentagons on to the first, one at a time. The last one can be tricky. But the model is very satisfying to build.

A key idea is to be able to make an accurate pentagon by drawing. This was described earlier (in Exercise 1) but will be described again recognizing its relation to the Golden section or Golden rectangle.

### Exercise 13   Pentagon construction

Draw a circle (red). Draw a square. Bisect the right-hand side. Draw radius to top left. Produce right-hand side upwards.

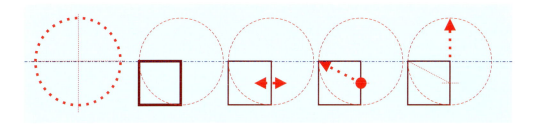

Fig 3.56

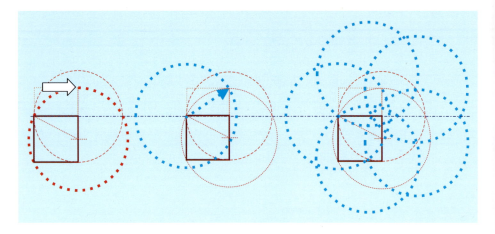

*Fig 3.57*

Draw an arc to arrow. Draw blue circle, centre left-hand top. Draw four more circles based on the initial red circle, this gives the five points of the pentagon

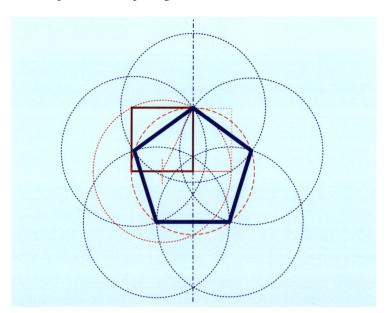

*Fig 3.58*

Find these five points and join them up. This gives the blue pentagon required.

## The thirteenth book of Euclid

It is interesting that the dodecahedron is almost the last to be dealt with in the last of the books by the great geometer Euclid.

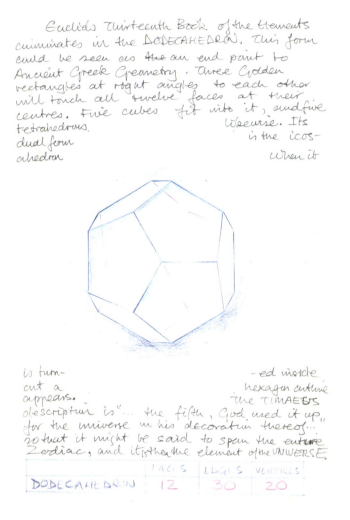

Euclids Thirteenth Book of the Elements culminates in the DODECAHEDRON, This form could be seen as the an end point to Ancient Greek Geometry. Three Golden rectangles at right angles to each other will touch all twelve faces at their centres. Five cubes fit into it, and five tetrahedrons. Its dual form is the icos- ahedron

When it is turn- -ed inside out a hexagon outline appears. The TIMAEUS description is "... the fifth, God used it up, for the universe in his decoration thereof..." so that it might be said to span the entire Zodiac, and it is then the element of the UNIVERSE

| | FACES | EDGES | VERTICES |
|---|---|---|---|
| DODECAHEDRON | 12 | 30 | 20 |

Fig 3.59  Dodecahedron construction

We see in Sir Thomas Heath's famous text in 1926 on Euclid's Elements that in Proposition 17 Euclid desires 'To construct a dodecahedron and comprehend it in a sphere ... and to prove that the side of the dodecahedron is the irrational straight line called apotome.' This process takes several pages. The drawing above (Fig 3.59) is nicely described in Robert Lawler's *Sacred Geometry*. In this text here it is sufficient to be familiar with the form as such — and be able to model it, the deductive details can be explored in a high school context.

The above is mentioned to show that such forms as the dodecahedron have been of interest to thoughtful communities more than 2 300 years ago. *Why* there was this interest is difficult to know, but modern studies do find that considerations of these solids is indeed necessary. From the form of viruses (tiny things), bucky-balls, Buckminster Fuller domes (big things), to the very shape of space itself (huge things).

## Euler's Law

Euler's Law is usually written as $F + V = E + 2$ which, being interpreted means that if we add the number of Faces to the number of Vertices this will equal the number of Edges plus two. He showed that this applied not only to regular figures but also irregular so long as they were convex.

I have rewritten this in the table so that

$$F - E + V = 2$$

*Exercise 14　Euler's Law for convex solids*

a) Check that the formula is true for all the Platonic solids. You should get *two* every time.

|  | Faces | Edges | Vertices |  |
|---|---|---|---|---|
| Tetrahedron | 4 | 6 | 4 |  |
| Octahedron | 8 | 12 | 6 |  |
| Hexahedron | 6 | 12 | 8 |  |
| Icosahedron | 20 | 30 | 12 |  |
| Dodecahedron | 12 | 30 | 20 |  |

b) Check to see if this law applies to the non-regular figures shown below.

i) YES / NO Yes

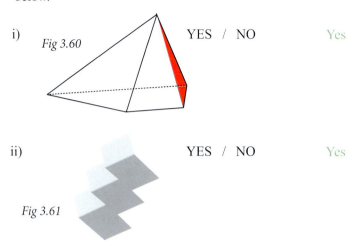

*Fig 3.60*

ii) YES / NO Yes

*Fig 3.61*

## Student work

The students often do magnificent work. Some samples of full sets of the five Platonic solids are shown here to close this

*Fig 3.62 Classwork*

*Fig 3.63 and 3.64  Classwork*

chapter. This is all work from Ruth P's Year 8 class of 2005 at Glenaeon Rudolf Steiner School.

# 4. Rhythms and Cycles

## Rotation and rhythm and cycles

In everything living there is some kind of rhythm. For the living has to die, and it reappears again with the fructification of seed. Rudolf Steiner mentioned that 'the comprehension of nature's rhythms will become true natural science.' This poses an enormous research challenge for any teacher (and scientist) to bring something contemporary to the students. This work can be a kind of preparation for changes in students of this age with their enormous psychic and hormonal swings.

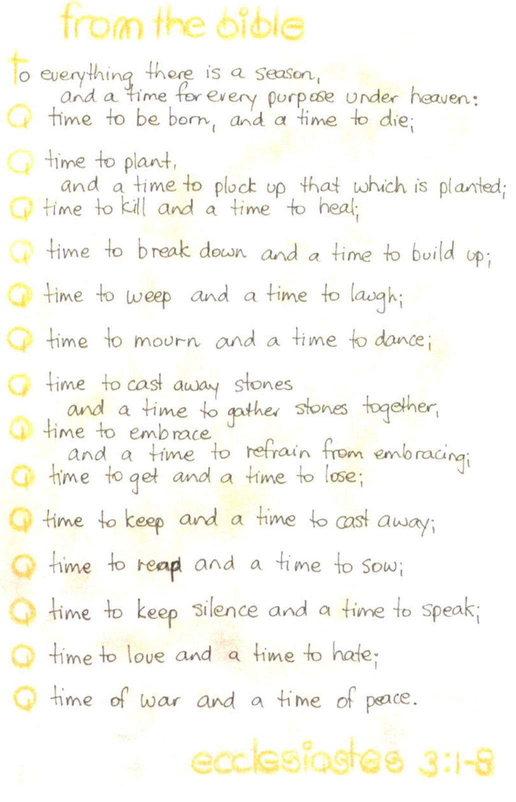

from the bible

To everything there is a season,
and a time for every purpose under heaven:
a time to be born, and a time to die;

a time to plant,
and a time to pluck up that which is planted;
a time to kill and a time to heal;

a time to break down and a time to build up;

a time to weep and a time to laugh;

a time to mourn and a time to dance;

a time to cast away stones
and a time to gather stones together;
a time to embrace
and a time to refrain from embracing;
a time to get and a time to lose;

a time to keep and a time to cast away;

a time to reap and a time to sow;

a time to keep silence and a time to speak;

a time to love and a time to hate;

a time of war and a time of peace.

ecclesiastes 3:1-8

Fig 4.1 'A Time to ....'

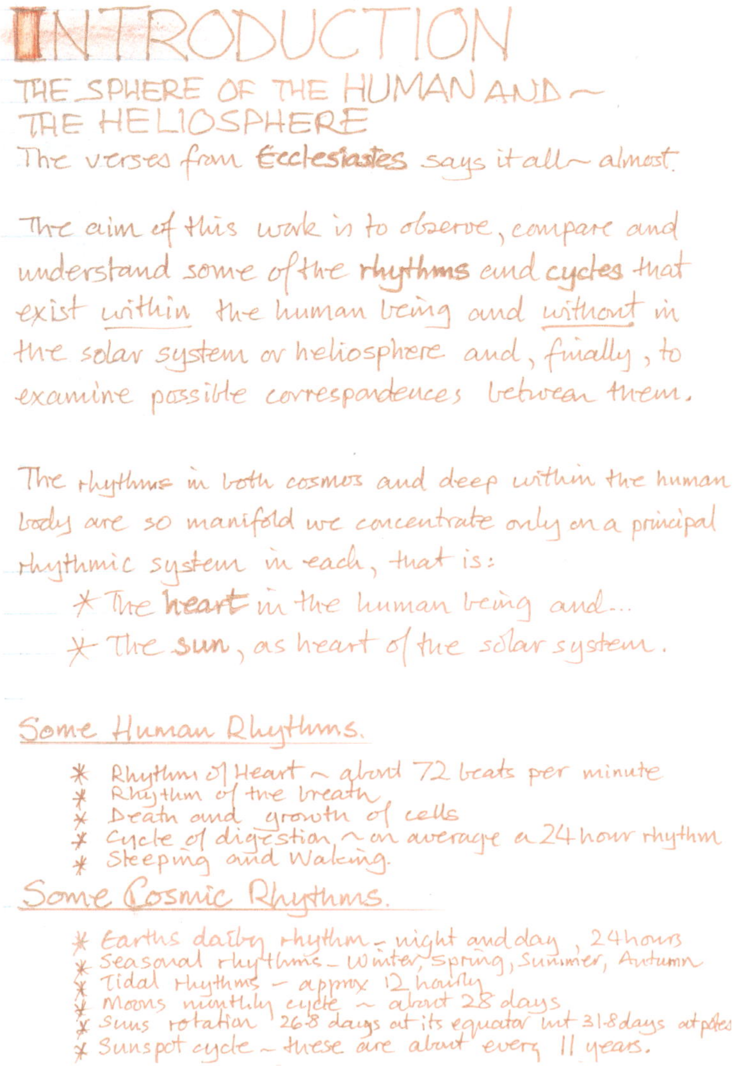

# INTRODUCTION
## THE SPHERE OF THE HUMAN AND ~ THE HELIOSPHERE

The verses from **Ecclesiastes** says it all ~ almost.

The aim of this work is to observe, compare and understand some of the **rhythms** and **cycles** that exist within the human being and without in the solar system or heliosphere and, finally, to examine possible correspondences between them.

The rhythms in both cosmos and deep within the human body are so manifold we concentrate only on a principal rhythmic system in each, that is:

    * The **heart** in the human being and...
    * The **sun**, as heart of the solar system.

### Some Human Rhythms.

* Rhythm of Heart ~ about 72 beats per minute
* Rhythm of the breath
* Death and growth of cells
* Cycle of digestion ~ on average a 24 hour rhythm
* Sleeping and Waking.

### Some Cosmic Rhythms.

* Earths daily rhythm - night and day , 24 hours
* Seasonal rhythms - winter, spring, summer, autumn
* Tidal rhythms - approx 12 hourly
* Moons monthly cycle ~ about 28 days
* Suns rotation 26·8 days at its equator but 31·8 days at poles
* Sunspot cycle ~ these are about every 11 years.

*Fig 4.2  Introductory — sun and heart*


## Time

That there is a sense of time, a time for *this* and a time for *that,* as put forward so evocatively in Ecclesiastes (3:1–8), is a picture of two sides to our world in a whole range of arenas. This duality is a further expression of the qualitative nature of two-ness, or duality. So it is sung in Ecclesiastes ... and the New Age musical 'Hair'!

*Fig 4.3 A main lesson title page 'Rhythm and Cycles, Life and Time' —
a student's work*

The dimension now added is that, not only are there poles that
contrast qualities but there is a swing or rhythm *between* the two.
This 'between,' which adds a special richness as the *crossover*
between two extremes is often a special time. Such as the crossover between night and day, day and night, or going to sleep and
awakening. The moments of the equinoxes, for instance, spring
and autumn, are special times leading to a solstice, itself a time
of some extreme.

There is an equivalence here to the cycle in the human being.
We breathe in and breathe out. We take in from the wider spheres

and give out from our being. Between these two is the *act* of breathing in and the *act* of breathing out. That is an underlying theme throughout this main lesson. The cycle of the seasons reflects one rhythm of the macrocosmos. The cycle of the breathing reflects one rhythm of the microcosmos.

   A simple spatial beginning can be made with the *circle*. And a beginning in time made with uninterrupted *rotation*.

## Wheels

Wheels rotate. They cycle round and round. It is what they do. If the cart's wheel was to do something different with each cycle we would get a trifle nervous. We expect a constant repetition of the same activity. This is perhaps the simplest cycle we can imagine. We talk of rpm and we mean revolutions per minute, each *revolution* or cycle being the same as the last.

   As soon as there is a difficulty in smooth untrammeled rotation, there is instant attentiveness. The squeaky wheel gets the oil. I remember as a teenager going straight over the handlebars of the pushbike when a wheel kind of jammed. Very salutary

*Fig 4.4  Model of a cart wheel*

it was. Especially when there is a crowd of peers watching! Another acquaintance got a stick in his front wheel. Instant stoppage. He had to be airlifted out and needed plastic surgery. It *does* matter that wheels continue to go round when expected to. This kind of cycling is always the same: repetitious, even boring, but we rely on it.

But there are all sorts of things that can happen during the cycles of very many phenomena in our solar system and many living systems. We will explore some later. What can be said about the circle and the circular directly?

## Circle and Diameter

The relation of a circle's diameter to its circumference. This ratio we know as π has puzzled many.

This has led to much exploration, even whole books, such as *A Biography of the World's Most Mysterious Number* by Posamentier and Lehmann, and *The Joy of π* by David Blatner.

Circles — how do we discover radius, diameter, perimeter and area? Is the circle the most perfect and yet simple form? This figure also led to one of the famous riddles from ancient Greek times — how to *square* the circle. What was meant here? Could one construct a square of the same area as a given circle

*Fig 4.5 Circle, pure and simple — or not so simple?*

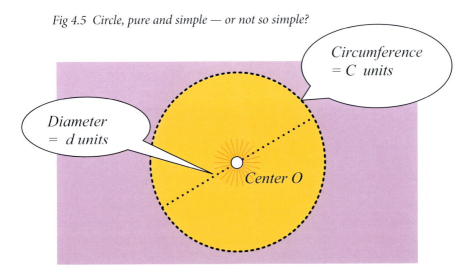

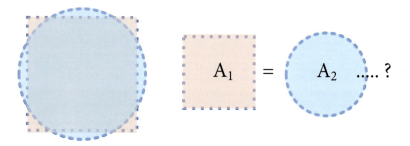

*Fig 4.6 Are the areas of the figures equal? With ruler and compass can we construct a circle of a given area the same area as a square. The answer is ... No!*

with only compass and ruler? Apparently it is impossible, and was proved to be impossible only in the nineteenth century by Lindemann. Is $A_1 = A_2$ in Fig 4.6?

What we see more readily of the circle is twofold. Later we see it is really threefold — at the very least. For we can ask, if the circle has a centre (and it does, for that is where we put the compass point to draw it!) then does it have a periphery? If it has an *inside,* has it an *outside?*

If the circle is imagined as the central cross section of a sphere (Fig 4.7) then we can also ask about the inside and outside of a sphere. A clue lies in shrinking the sphere indefinitely — until, so we imagine — it becomes a *point.* Now imagine infinite enlargement. To what does the sphere tend? It gets flatter and flatter. This is in both directions, to the left and to the right in the narrow band. Take this to the infinite, no, not *near* the infinite, but *to* the infinite and it becomes flat. Yet it is the same sphere. The notion of polarity begins to loom. The sphere itself (whatever its size) lives between the two poles of *point, O,* and (say) *plane* o. And this is as much as I would draw attention to here. In the Platonic solids main lesson polarity emerges there too with the polar nature of the five solids.

It would look as if the circle then would tend to a line at infinity. This is worth trying to think through, but students at this time do not like this apparent paradox. Nevertheless, our sphere meets us everywhere too. In Sun, Moon, Venus and bubbles (Fig 4.8)!

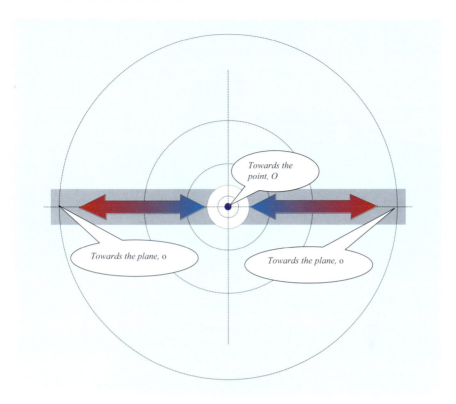

▲ *Fig 4.7 Expanding and contracting sphere*

*Fig 4.8 Rachel and spheres galore* ➤

For the moment it is enough to explore briefly what whole books have been written on. For a given circle (which means both circumference and diameter are married to each other in a fixed relationship) what *is* this relationship in terms of the two of them?

## Circumference and diameter

To get a class estimate of this relationship is quite straightforward. It is also a useful and practical exercise, in terms of careful measurement and then simple statistics for the whole group.

*Exercise 15   Diameter and circumference, an empirical approach*

1. Find a round tin, bottle, or jam jar (it is best if the ends are flat). and a length of thin string about 40 cm for convenience).

2. Tape one end of the string securely to the middle of the tin so that the line of the string is at right-angles to the axis of the tin. Mark a point (green dot) at tape and string junction.

*Fig 4.9*

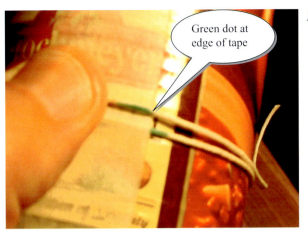

*Fig 4.10*

Green dot at edge of tape

3. Wind the string all the way round the tin and overlap the place on the tape where the first mark was placed. Mark this second point.

4. Unwind the string and take off the tape.

5. Carefully measure the length between the marks on the tape. This will be the circumference, *C*, of the tin.

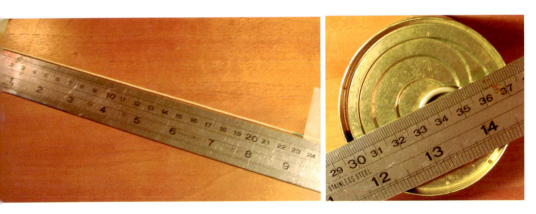

*Figs 4.11 and 12*

6. Carefully measure the diameter, *d*, of the tin. (Take account of the overlapping edge).

7. Record these measurements in a table such as below. Perform the division sum indicated (that is, divide the circumference,

*C*, by the diameter, *d*) and record this value. It is good if the class has different size circular objects to get a range of results.

| Name | Circumference (*C* cm) | Diameter (*d* cm) | *C* ÷ *d* = *k* |
|---|---|---|---|
| | | | |
| JB | 23.2 cm | 7.3 cm | 23.2 ÷ 7.3 = 3.18 |
| | | | |
| | | | |
| | | | |
| | | | |
| | | | |

In this case my answer is that *k* = 3.18. In other words the circumference is about 3.18 times the diameter. This is about right from all accounts! Are there other ways of finding out what this value of *k* is? Archimedes thought so. He lived about 300 BC and was born in Sicily. He calculated limits between which this figure, *k*, must lie.

## Archimedes' approach with polygons

He took a geometrical approach — sometimes called a *method of exhaustion*. An exercise for some of the simpler cases is outlined below.

*Exercise 16   Rough estimate for diameter and circumference relationship with squares*

Assume a circle of 10 cm diameter and find the perimeter of a polygon circumscribing the circle and a polygon inscribing the circle. Let this polygon be a square. Then for unit diameter the circumference of the circle must lie between the length of these two polygons.

1. Draw a circle of 10 cm diameter on centre *O*.

2. Draw a square around the circle and within the circle as shown and add diagonals.

3. Measure the distances on each side of the two squares. If constructed accurately then the outside square should have a total perimeter of close to 40 cm. The inside square should be close to 28 cm.

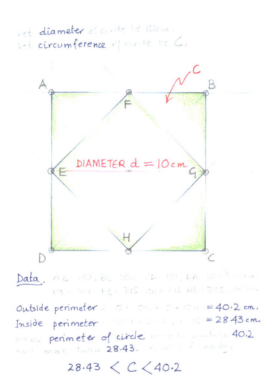

*Fig 4.13*

28·43 < *C* < 40·2

4. Assume the perimeter length of the circle, its circumference, is *C* units. Then we can write (as in the drawing):

$$28.43 < (C \times 10) < 40.2$$

Now divide by 10 all the way through (so that the diameter is 10/10 = 1 unit):

$$\frac{28.43}{10} < \frac{(C \times 10)}{10} < \frac{40.2}{10}$$

And we get:

$$2.843 < \quad C \quad < 4.02 \text{ when diameter} = 1 \text{ unit}$$

So we know our magic number lies between 2.843 and 4.02. This is not particularly helpful for as even empirically we got 3.18 with a piece of string and a soup tin. Can we do better? Archimedes used this multi-polygon method for many more polygons, going up to, apparently, a 96-sided figure both inside and out.

*Exercise 17    A less rough estimate for diameter and circumfer-
ence relationship but with hexagons inside and out*

1. Draw circle where $d = 10$ cm

2. Mark off equal radii $A$ to $F$ as shown.

3. Bisect angles $AOB$, $BOC$ etc and put in inner hexagon as shown.

Fig 4.14

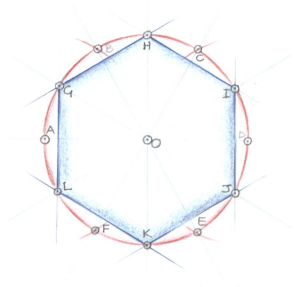

4. Determine a tangent to the circle at G by
drawing a perpendicular to line GOJ, a diameter.

*Fig 4.15*

5. This tangent line cuts BOE at a point N and
AOD in M. Set compass at radius OM and draw a
circle at center O.

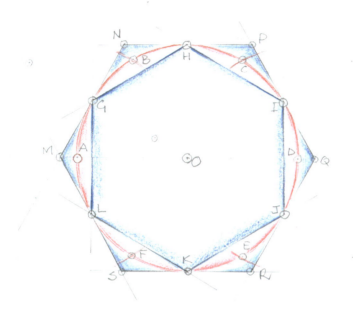

4. Determine a tangent to the circle at *G* by drawing a *perpendicular* to line *GOJ*, a diameter.

5. This tangent line cuts *BOE* at a point *N* and *AOD* in *M*. Set compass at radius *OM* and draw a circle centre *O*.

6. Join *M* to *N*. This is one side of the external hexagon. Draw in the remaining points *P, Q, R* and *S*.

7. Now complete the further five sides of this external hexagon.

8. Measure the outside lengths and add to get total *external* perimeter of hexagon. Measure the inside lengths and add to get total *internal* perimeter of hexagon.

9. Write these values down as follows:
   Internal Perimeter < (*C* × 10) < External Perimeter

Another way is to *calculate* and this is possible in this relatively simple case. So we can check our measurement with calculated values (in this case).

## Exercise 18   *Internal and external hexagons, by calculation*

1. Abstract the triangle *OIJ*. Note that the radius of the circle, centre *O*, is half the diameter, that is 10/2 = 5 cm.

2. Since triangle *OIJ* is equilateral (through its very construction with compass set at equal radii), then the vertical length *IJ* is also  5 cm

3. But *IJ* is one of the sides of the *internal* hexagon, so the perimeter of this internal hexagon must be equal to:
   6 × 5 cm = 30 cm

4. The perimeter of the external hexagon is a little more complicated to calculate. Draw the perpendicular to *IJ* through *O* meeting *IJ* in *T*.

5. Now form the triangle *OQI* where angle *OIQ* is a right-angle. Note that *IT* is half of *IJ*, i.e. 5/2 = 2.5 cm.

*Fig 4.16*

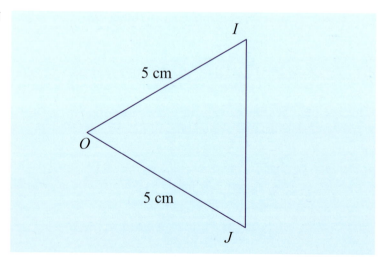

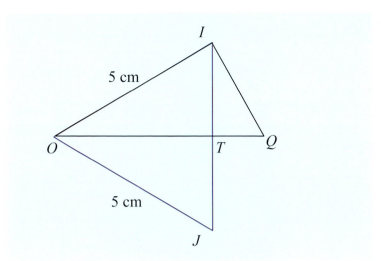

*Fig 4.17*

6. Calculate length of *OT* using Pythagoras' Theorem $a^2 + b^2 = c^2$.

$OT^2 + IT^2 = OI^2$ or $OT^2 + 2.5^2 = 5^2$

Hence $25 - 6.25 = OT^2$, hence $18.75 = OT^2$ and hence $4.3301 = OT$

7. Now to find *IQ*. This depends on using the two *similar* triangles $\triangle OIT$ and $\triangle OQI$. Since they are similar their *corresponding* sides are in the same ratio, that is $IQ / IT = OI / OT$, and substituting the known values, $IQ / 2.5 = 5 / 4.3301$

Therefore $= 2.5 \times (5 / 4.3301)$, hence $IQ = 2.8867$.

8. But *IQ* is half of *PQ*, and *PQ* is a side length of the external hexagon.

So $PQ = 2 \times 2.8867$ Hence $PQ = 5.7734$.

9. This means that the external hexagon is $6 \times 5.7734 = 34.640$ in length.

10. In conclusion: $30 < (C \times 10) < 34.640$ and divide all through by 10 giving

$3 < C < 3.4640$

So when $d = 1$, *C* is between 3 and 3.464.

## *And with eight sides ....*

Finally, for eight-sided polygons (Fig 4.18), by measurement we get:

$$30.4 < (\pi \times 10) < 33.2$$

(for $\pi$ is the symbol given to *C* in this case) so that:

$$3.04 < \pi < 3.32$$

It is said that Archimedes found using a 96-sided polygon that *C* must lie between 3 and one seventh and 3 and ten seventy firsts. This was quite an achievement and more than accurate enough for most purposes. If we take the average of these two values:

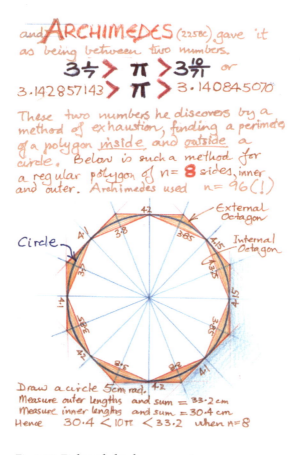

and ARCHIMEDES (225BC) gave it as being between two numbers.

$$3\tfrac{1}{7} > \pi > 3\tfrac{10}{71} \quad or$$

$$3.142857143 > \pi > 3.140845070$$

These two numbers he discovers by a method of exhaustion, finding a perimeter of a polygon *inside* and *outside* a circle. Below is such a method for a regular polygon of n= **8** sides, inner and outer. Archimedes used n= 96 (!)

Circle

External Octagon

Internal Octagon

4·2    3·8    3·85    4·15

Draw a circle 5cm rad.
Measure outer lengths and sum = 33·2 cm
Measure inner lengths and sum = 30·4 cm
Hence    30·4 < 10π < 33·2  when n=8

*Fig 4.18 Eight-sided polygon, or octagon*

$3^1/_7 = 3.1428571428571$ which repeats after the seventh digit

and $3^{10}/_{71} = 3.1408450704225$

which is $(3.1428571428571 + 3.1408450704225) \div 2 = 3.1456539235411$ and this is accurate to two decimal places only, that is 3.14 — which only goes to show what a lot of work $\pi$ can be.

## *Naming of $\pi$*

The ratio of circumference to diameter was not named as such until the early 1700s (Denis Guedj 1996, 100). It was given the symbol $\pi$, the first Greek letter (usually pronounced 'pie') of the Greek word 'periphery' (Gullberg 1997, 85).

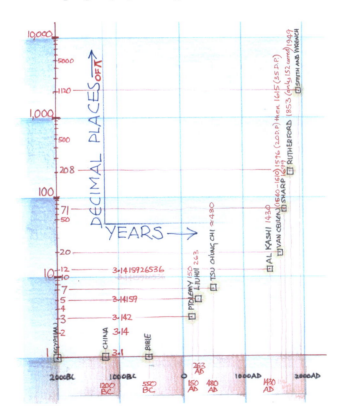

*Fig 4.19 What has $\pi$ been recorded as over the last four millennia? (Data from Posamentier & Lehmann 2004, 75f)*

It was 'in 1706 the English mathematician William Jones ... used the symbol π for the first time to actually represent the ratio of the circumference of a circle to its diameter' (Posamentier & Lehmann 2004, 67) and following this was the popularization due to the famous Euler in 1748. It needed a name as until 1706 'one had to content oneself with the quaint Latin phrase *quantitatis, in quam cum mulitplicetur diameter, provenient circumferentia* meaning "the quantity which, when the diameter is multiplied by it, gives the circumference".' So *pi* was a useful shorthand.

In the early 1600s it had been called the Ludolphian number after Ludolph van Ceulen who found π to 20 decimal places.

## *Increasing accuracy over time*

Its value is now known to a very great number of decimal places but its most common approximation is 3.142 or 3.1416. The drawing in Fig 4.19 (p.179) gives a small selection of those who have tried to determine this ratio which we now call π.

Many approximations have come down over the ages and only a few have been shown here. Even in the Old Testament of the commonly received Bible there is a story that suggests this ratio: 'Then he made a molten sea; it was round ten cubits from brim to brim and five cubits high, and a line of thirty cubits measured its circumference.' (1Kings 7:23)

This looks suspiciously like a statement about a circle with a 10 cubit *diameter* and a 30 cubit *circumference*. This makes us believe that the writer of Kings assumed π was of the order of three. This was a pretty big bowl he was talking about if we believe the cubit is a unit of measure the length of the forearm.

The chart in Fig 4.19 does not attempt to cover the advances taken today as, with current computers, people have used machine and algorithm to calculate the ratio π to *billions* of decimal places. I cannot be sure of the significance of this! It seems that no pattern has been discerned in the numbers following the '3,' yet if the thing can be calculated there has to be the pattern contained in the algorithm itself. Some examples of such series patterns as this are listed by Blattner, Gullberg & Posamentier where π is:

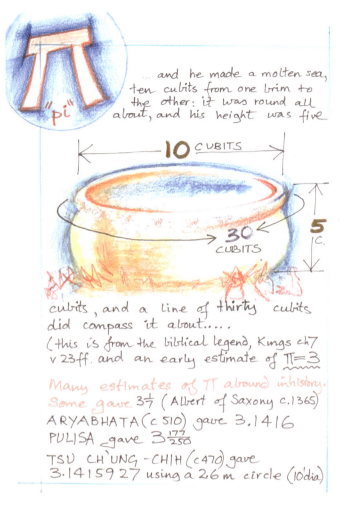

... and he made a molten sea, ten cubits from one brim to the other: it was round all about, and his height was five

cubits, and a line of thirty cubits did compass it about.....

(this is from the biblical legend, Kings ch7 v 23ff. and an early estimate of π=3

Many estimates of π abound in history. Some gave 3⅐ (Albert of Saxony c.1365) ARYABHATA (c 510) gave 3.1416 PULISA gave 3 177/250 TSU CH'UNG-CHIH (c470) gave 3.1415927 using a 2.6 m circle (10'dia)

*Fig 4.20 'and a line of thirty cubits measured its circumference'*

$$\pi = \cfrac{2}{\sqrt{\dfrac{1}{2}} \times \sqrt{\dfrac{1}{2}+\dfrac{1}{2}\sqrt{\dfrac{1}{2}}} \times \sqrt{\dfrac{......}{......}}}$$   due to Viète in 1593

$$\pi = 2 \times \left( \frac{2\times2\times4\times4\times6\times6\times8\times8....}{1\times3\times3\times5\times5\times7\times7\times9....} \right)$$   due to Wallis in 1655

$$\pi = 4\left( 1-\frac{1}{3}+\frac{1}{5}-\frac{1}{7}+\frac{1}{9}-...... \right)$$   due to Gregory (1670), and a later Leibniz (1673)

It is extremely intriguing that such a variety of different series and formulae give us the *same* number. Mentions have been made since early Grecian times that there may be something to the notion that such a number, with so many ways of finding it, yet no discernible pattern, is a link between the curvaceous and the linear, between the divine and the human, between the circular and the square. This has puzzled the human community since at least those of ancient Greece tried to square the circle.

## Circumference of a circle

Having agonized over this special value and relationship we could check again in empirical terms. Some examples are listed for various round objects in Fig 4.21.

If the relationship is such that when the diameter is 1 unit and then the circumference is $\pi$ units then we can finally write:

Diameter    :   Circumference

1        :        $\pi$

multiply both sides by $d$

$d$        :        $\pi \times d$

But $d = 2 \times r$ so therefore:

2 × $r$      :      $2 \times \pi \times r$

Or as it is usually expressed

$$C = \pi \times 2 \times r$$

or more simply:

$C = 2\pi r$, the circumference, $C$, of a circle with radius, $r$ units.

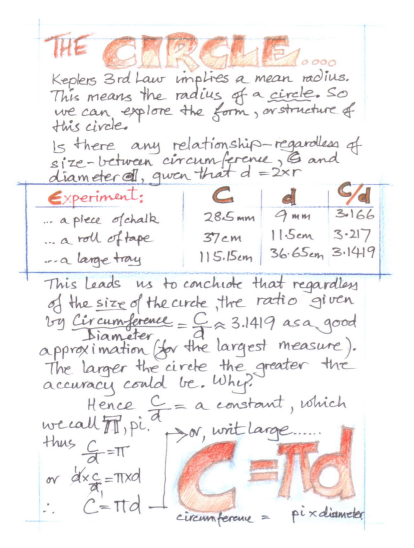

**THE CIRCLE** o ooo

Keplers 3rd Law implies a mean radius.
This means the radius of a circle. So
we can explore the form, or structure of
this circle.

Is there any relationship—regardless of
size—between circumference, C and
diameter d, given that $d = 2 \times r$

Experiment:

| | C | d | C/d |
|---|---|---|---|
| ... a piece of chalk | 28.5 mm | 9 mm | 3.166 |
| ... a roll of tape | 37 cm | 11.5 cm | 3.217 |
| ... a large tray | 115.15 cm | 36.65 cm | 3.1419 |

This leads us to conclude that regardless
of the size of the circle, the ratio given
by $\dfrac{Circumference}{Diameter} = \dfrac{C}{d} \approx 3.1419$ as a good
approximation (for the largest measure).
The larger the circle the greater the
accuracy could be. Why?

Hence $\dfrac{C}{d} = $ a constant, which
we call $\pi$, pi. $\longrightarrow$ or, writ Large......
thus $\dfrac{C}{d} = \pi$

or $\overset{1}{d} \times \dfrac{c}{d_1} = \pi \times d$

$\therefore \quad C = \pi d$

$$C = \pi d$$

circumference = pi × diameter

*Fig 4.21 Circumference of circle*

Which is a little formulation that is worth remembering but is often confused with $\pi r^2$ by the students — or so I have found. This other little formulation happens to be for the area of a circle so we will deal with this now. It is not surprising that with such a curious number, if the relation of periphery and diameter (or radius) causes such a fuss, then there is no reason to believe that finding the area will be any less fraught.

We start with a neat exercise, using cut-out cardboard.

*Exercise 19   Finding an approximation to the area of a circle*

This would be extraordinarily easy if we could *square* the circle. But we can't (and it has been proved we can't when Lindemann showed the transcendence of $\pi$). Transcendence means here that it cannot be the root, or solution, to any equation built of ordinary rational numbers and a limited number of terms.

Take a circle, cut it up, rearrange it into an approximate rectangle, and calculate. Easy. This is outlined in Fig 4.22.

1. Draw a circle of 4 cm radius.

2. Divide the circle into 16 equal parts (i.e. divide circle into four, halve these quarters, then halve again, using the bisection of an angle construction elaborated in Class 7). Each sector should be 22.5° at the pointy end.

3. Rearrange these 16 pieces so that they form something like a parallelogram as shown.

4. This parallelogram will have an approximate height of $r$ cm (in this case, 4 cm). It will also have a base of about half the circle's circumference or $C = 2\pi r / 2, = \pi r$.

5. As the area of a parallelogram is $A$ = base × vertical height, $= b \times h$ and since $b = \pi r$. and $h = r$ roughly, then $A \cong \pi r \times r$, or we say:

Area of circle is $A \cong \pi r^2$.
   ($\cong$ means approximately equals to)
   So the area of this particular circle is $A \cong \pi \times 4^2$, or $A \cong \pi \times 16$, $A \cong 16\pi$ square cm.

If we were capable of making very narrow sectors, and then do the above we would tend to an exact rectangle and the area *would* be: $A = \pi r^2$.

$$A = \pi r^2 \text{ where } A \text{ is the area and } r \text{ is the radius}$$

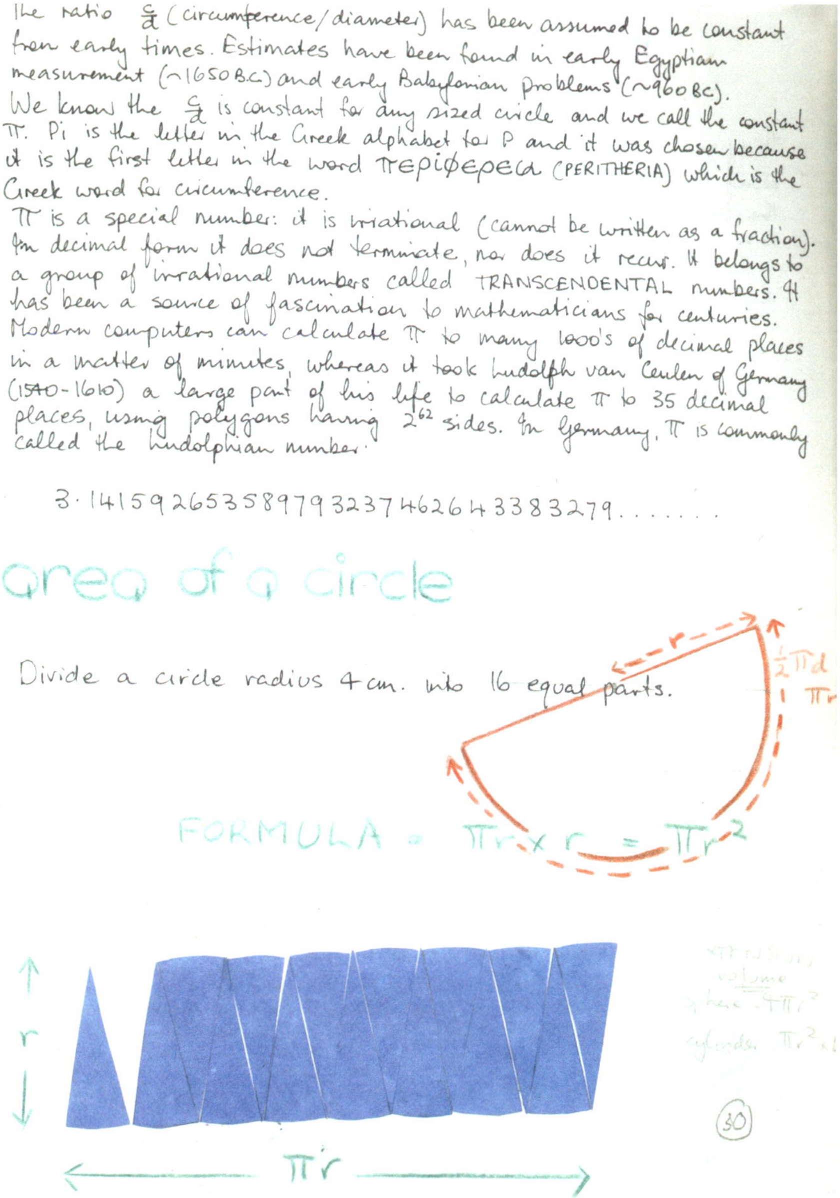

4. RHYTHMS AND CYCLES

The ratio $\frac{c}{d}$ (circumference/diameter) has been assumed to be constant from early times. Estimates have been found in early Egyptian measurement (~1650 BC) and early Babylonian problems (~960 BC).

We know the $\frac{c}{d}$ is constant for any sized circle and we call the constant $\pi$. Pi is the letter in the Greek alphabet for P and it was chosen because it is the first letter in the word ΠΕΡΙΦΕΡΕΙΑ (PERITHERIA) which is the Greek word for circumference.

$\pi$ is a special number: it is irrational (cannot be written as a fraction). In decimal form it does not terminate, nor does it recur. It belongs to a group of irrational numbers called TRANSCENDENTAL numbers. It has been a source of fascination to mathematicians for centuries. Modern computers can calculate $\pi$ to many 1000's of decimal places in a matter of minutes, whereas it took Ludolph van Ceulen of Germany (1540-1610) a large part of his life to calculate $\pi$ to 35 decimal places, using polygons having $2^{62}$ sides. In Germany, $\pi$ is commonly called the Ludolphian number.

$$3.14159265358979323746264 3383279\ldots\ldots$$

## area of a circle

Divide a circle radius 4 cm. into 16 equal parts.

$$\text{FORMULA} = \pi r \times r = \pi r^2$$

Fig 4.22  Area of a circle by reconstituting the dismembered circle as an approximate rectangle — drawing and notes by Anne Jacobsen

## Micro, meso and macro scales

The wheel is on, so to say, the *meso*scale. It is somewhere in the middle of the *micro* and the *macro*.

At the microcosmic level we have the tiny cycles of the wavelengths of light that are used in the modern world to define the length of the metre. There are 299 792 458 such oscillations every single second. Hardly a practical definition for carpenters! But the physical scientist needs this to do a particular kind of investigation. This seems not very relevant when it is realized that there is very roughly one heart beat every second. That level of measure we can sense. That has a human perspective.

At the macrocosmic level we have the mighty cycles represented, we are told, by the majestically slow rotation of the mighty spiral galaxies. The Sun it is said moves through the galaxy and Hoyle (1962, 257) states the Sun's velocity to be 150 miles per second (or 240 km/sec) through this galaxy. The numbers here also get beyond our comprehension.

So cycle and rhythm are at all kinds of *scales* and vastly different measures of *speed*. And at the very core of music is rhythm, or beat (sometimes to the exclusion of almost all else if one listens to the noise emanating from the boom boxes in many a car!). These are physical rhythms. It has been suggested that if one wants to know the secret of life then this is *also* rhythm but at quite another level. Here the rhythm is quite different to mechanical beat and brings about distinction, qualitative difference, during any cycle — not mere repetition. Variation occurs throughout the cycle. Imagine the ellipse. Even here we can imagine a speed change as we traverse its path — like the planets' movement as Kepler discovered, with rapid speed at one end of the ellipse and slower at the other. But now to revert to the simple.

## The circle

The circle, or the *circular*, could be thought of as the form in *space* of the wheel's revolving in *time:* Circles in space; cycles

*Fig 4.23 Venus in front of the sun. Erik T (past student, now with CSIRO) projecting the image of Venus on to a surface.*

*Fig 4.24* ➤

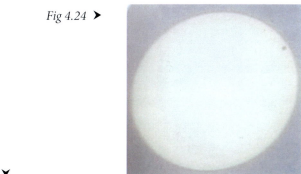

*Fig 4.25* ▼

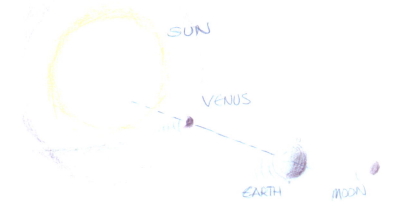

in time. We see the circular whenever we inadvertently glance at the sun. Easier to see is the lunar circle.

It was this event that Captain Cook had orders to observe in the Pacific all those years ago in the late 1700s. Infrequent? Yes, but, nevertheless a regular and cyclic event. I think it is important that students see these things. They remind us of the great rhythms in which we are embedded (Fig 4.26).

# Venus transit

### By  JB    Transit Reporter

If the Sun is bright today and there is little cloud then we should be able to project an image of the Sun on to a screen and watch the planet Venus (about $^1/_{30}$ the size of the Sun) transit from 3.07 pm until nearly sundown, today, Tuesday, 8th June 2004.

Do not look at the Sun itself and certainly not through any optical device.

Venus transits this year over the upper part of the Sun's disc (see panel below). It appears as a black dot slowly traversing the Sun. Starting to touch the Sun at 3.07 pm it is almost another 20 minutes before it is fully on the Sun. It is a real slow coach compared to Mercury. The next transit will be in eight year's time, 2012. Before this year it has not been seen since 1882. If you are in Class 6 now you will have left school before it happens again.

*Fig 4.26*

The length of time that the planet Venus takes to do its full cycle is something that we do not need any cosmological theories to determine, at least with reference to the Earth. We simply watch the sky. For Venus has a cycle that observers for thousands of years have taken note of. This cycle translates into 0.615 earth years (224.701 / 365.256 = 0.615). We know the earth's cycle also through actual observation — about 365.256 days — and this is empirical. This is seen through the time it takes for all the four seasons to come round again.

There are two things here: the circular discs in *space* of Sun and Venus and the cyclic rhythm of Venus transit in *time*.

## Day and night and the Nebra disc

Day follows night follows day and so on ... However the length of 'day' is never the same. Anyone living in temperate to high latitudes knows this well. I remember working in the south of England in winter as an apprentice and we would get to work in the dark and leave work in the dark. Common joke — if you blink you will miss the sun (on a cloudless day)! And this was only at 50 something degrees north. All this has to do with when and where the sun rises in the morning. The *range* of the horizon over which this rise occurs is quite remarkable and depends on latitude.

Just how remarkable it is, is shown by a sketch of a beautiful bronze-age artifact (see Fig 4.27) discovered not so long ago. A hoard discovered by treasure hunters in Eastern Europe came up with this intriguing artifact, which, it is claimed, represented a quite new slant on what the bronze-age peoples knew within their cultures. If it is no fake (its veracity has inevitably been challenged) then it suggests such people knew their astronomy. It has been called the Nebra sun-disc or sky-disc.

It is 32 cm in diameter. There are a number of interpretations. The left gold circle represents the sun (or day), and that the right figure represents the crescent moon (or night). The seven small circles may represent the stars of the Pleiades.* The arc below

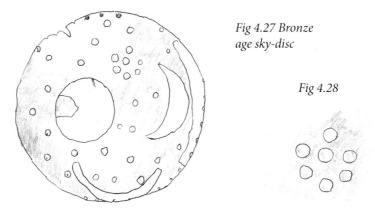

*Fig 4.27 Bronze age sky-disc*

*Fig 4.28*

---

* Mythology speaks of these stars being seven in number — seven gentle daughters of Atlas. But to the naked eye, usually only six Pleiades are visible. This has given rise to the notion that one of the group has dimmed or become lost. The 'lost Pleiad' figures in the mythologies of several cultures down the ages. (Davidson 1993, 17)

Fig 4.29

represents a 'sky boat' taking us from night to day and back again, a theme and symbol current in parallel cultures.

The arcs on edge at left (gold is missing) and at the right represent the *range* of sunrise and sunset over the year at the location the artifact is native to (see Fig 4.30).

This angle has been interpreted as the angle over which sunrise and sunset traverses at the particular latitude during the year. It may seem astonishing that the ancients (estimated about 2000 BC) should have known such things. This angle is 82° on the disc and corresponds to the range for the latitude at which the disc was found. And could the other small gold circles also be other stars? Speculation is rife!

This angular swing is due to the orientation of the Earth axis to its orbit which is inclined at 23.5°. This orientation of the

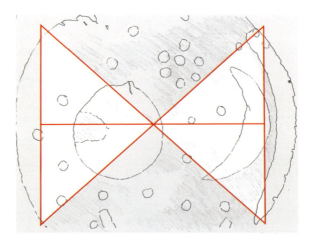

*Fig 4.30  Angle of both arcs is about 82 degrees. Their center could even be on the edge of the 'sun' image.*

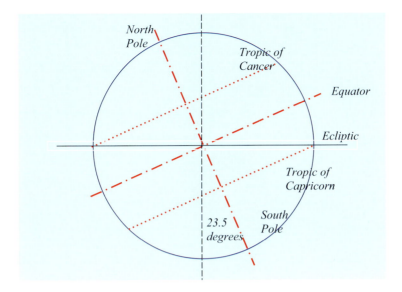

*Fig 4.31  Earth tilt and the Tropics*

north/south axis defines the *tropics,* the Tropic of Cancer in the northern hemisphere and the Tropic of Capricorn in the southern hemisphere. These are circles of latitude on the Earth's surface at 23.5° north and south of the equator.

It is because the sun appears to spiral up and down between the tropics that in summer the sun rises and sets to the north of east and west and in winter it rises and sets towards the south — in the northern hemisphere. In the southern hemisphere the seasons are opposite, and in summer sunrise and sunset are to the south of east and west, while in winter to the north. This is the angular swing which is claimed to be represented in the Sun disc. The raw mechanics of this change in Earth orientation are as follows.

## Current basic picture based on Copernicus' ideas

If the basic picture of our Earth is conceived as above then we see the poles tilted at an angle to the vertical (Fig 4.31). This angle is reported as 23.44 degrees.

This angle seems to be a constant over a very long time and it is because of this tilt in changing relation around the ecliptic that we have *seasonal* change.

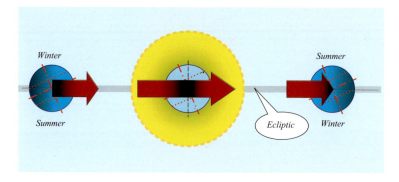

*Fig 4.32 Effect on Earth of tilt and the generation of the different seasons*

## Seasons

In Fig 4.32 we see the effect of this Earth axis tilt. On opposite sides of the orbit of the Earth around the Sun the seasons switch slowly over. If we imagine in summer the being of the Earth as breathing out, and in winter the Earth as breathing in, then at any one of these extreme times the Earth breathes *in,* and at the same time breathes *out* in the opposite hemisphere: 'out-breathes at one place and in-breathes at another' (Steiner 1923, *The cycle of the year,* pp.2f). So for the being of the Earth it is both winter and summer *at the same time* but in different hemispheres every half year.

This is complicated enough, but the Earth does other things. It is true, in this picture, that it always moves in the *ecliptic.* This is the plane of the Earth's orbit (we have to start somewhere!) that intersects with the Sun's centre and the Earth's motion around the Sun is the same as all the other planets. Usually this is regarded as anticlockwise when viewed from 'above,' or way above the North Pole. It is then called *direct.*

## Earth's elliptical path round the Sun

The path that the Earth moves in the ecliptic is an almost circular *ellipse* (it only has an eccentricity of 0.0167) (Fig 4.33). Even so this brings the Earth near and further away from the Sun.

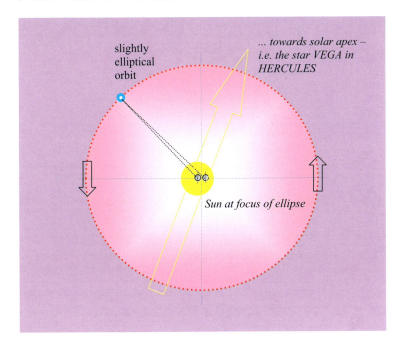

Fig 4.33 Very slightly elliptical path of the Earth around the Sun

The close approach of the Earth to the Sun is called the *perihelion* and the furthest distance is the *aphelion*. The perihelion point makes one eastward revolution round the zodiac in about 110 000 years. This is a slow movement indeed.

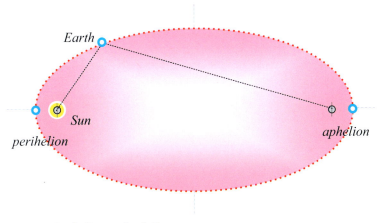

Fig 4.34 Perihelion and aphelion

Even the mighty Sun is said to be in movement as well
— just to complicate matters — towards one of the constellations. This is the constellation of Hercules and is near the star
Vega. This is known as the *solar apex.* The opposite direction
is the *solar antapex.* The speed towards Vega is said to be
19.7 km/sec.

For *descriptive purposes only* the exaggerated ellipse for
Earth orbit (Fig 4.34, p.193) can show more clearly the ideas of
perihelion and aphelion.

This all has to be some kind of cosmic music. In the cosmic
and in the solar system and the Earth, Moon and Sun systems
particularly, there are also the most intriguing relationships.

Such relationships are some of those which Johannes Kepler discovered between the planets. There are three of these in particular.

Above was mentioned that the path of the Earth with respect
to the Sun is an ellipse. Who was the individual who first had this
idea — after many years of work? It was Johannes Kepler (he
published the first law in 1609 and the last in 1619).

## Kepler's planetary laws

He discovered three principal relationships. If we squash a circle
in a special way we get another form. This form is called the
ellipse. What is so special about an ellipse? It is — if Kepler is

*Fig 4.35  Cycles and epicycles from the times of both Aristotle and Ptolemy*

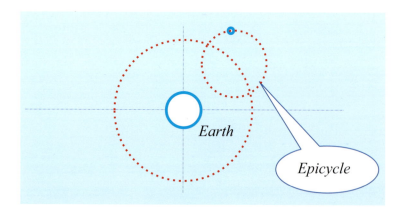

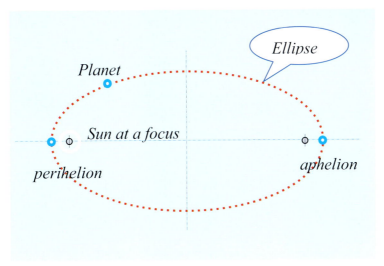

Fig 4.36

to be believed — the track of the very body, the Earth, that all of us are riding on! The path of the Earth about the Sun is said to be elliptical (see above).

In Greek times, Ptolemy and Aristotle had seen the planets moving around the Earth in circular paths. The planets though moved on an epicycle, a circle upon this circle (Fig 4.35). However, to accurately explain the irregularities of the movement small circles had to be added to the epicycles — a very complicated picture.

Then in 1453 Copernicus showed a greatly simplified picture by putting the Sun in the centre, and having all the planets move in circles around the Sun. While this was an enormous simplification, there remained inexplicable irregularities. It took Kepler's genuis to recognize the planets moved in *ellipses* with the centre of the Sun at one focus. *This is Kepler's first law.*

The *first* law he discovered was:

(1) That all the planets follow a path of an
    ellipse with the Sun at one focus.

*Fig 4.37*

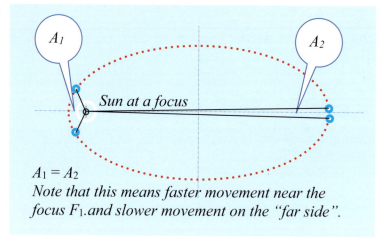

$A_1 = A_2$
*Note that this means faster movement near the focus $F_1$. and slower movement on the "far side".*

The *second* law (see Fig 4.37) he discovered was:

> (2) The radius vector from Sun to a Planet sweeps out equal areas in equal times.

The *third* law he discovered was:

> (3) The square of the orbital period, $T$, is directly proportional to the cube of the mean distance, $R$.

Put in symbolic terms this is also:   $T^2 = kR^3$

$$\frac{T^2}{R^3} = k$$

where $T$ is the orbital period and $R$ is the mean or average distance to the Sun from the Planet and where $k$ is some constant dependent on the units used for both time and distance.

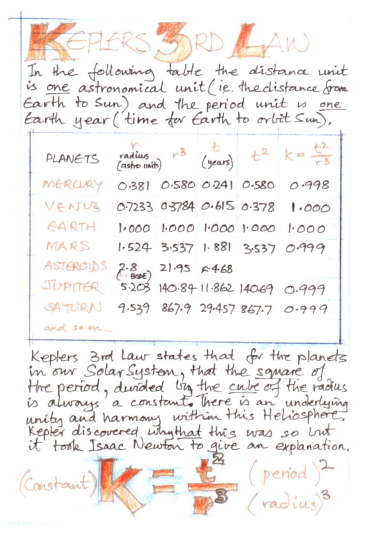

## KEPLERS 3RD LAW

In the following table the distance unit is __one__ astronomical unit (ie. the distance from Earth to Sun) and the period unit is __one__ Earth year (time for Earth to orbit Sun).

| PLANETS | r radius (astro units) | r³ | t (years) | t² | $k = \dfrac{t^2}{r^3}$ |
|---------|------------------------|-----|-----------|-----|-----------------------|
| MERCURY | 0.381 | 0.580 | 0.241 | 0.580 | 0.998 |
| VENUS | 0.7233 | 0.3784 | 0.615 | 0.378 | 1.000 |
| EARTH | 1.000 | 1.000 | 1.000 | 1.000 | 1.000 |
| MARS | 1.524 | 3.537 | 1.881 | 3.537 | 0.999 |
| ASTEROIDS | 2.8 (& BODE) | 21.95 | ≈468 | | |
| JUPITER | 5.203 | 140.84 | 11.862 | 140.69 | 0.999 |
| SATURN | 9.539 | 867.9 | 29.457 | 867.7 | 0.999 |

and so on...

Keplers 3rd Law states that for the planets in our Solar System, that the square of the period, divided by the cube of the radius is always a constant. There is an underlying unity and harmony within this Heliosphere. Kepler discovered why that this was so but it took Isaac Newton to give an explanation.

$$\text{(Constant)} \quad K = \frac{t^2}{r^3} \quad \frac{(\text{period})^2}{(\text{radius})^3}$$

*Fig 4.38  Kepler's third law*

An example may be of interest. Compare for instance the Earth's neighbours, Venus and Mars.

VENUS:    Sidereal period, $T_V = 224.701$ days
             Mean distance, $R_V = 108.21$ million km

MARS:    Sidereal period, $T_M = 686.980$ days
             Mean distance, $R_M = 227.94$ million km

So is it true that $\dfrac{T_V^2}{R_V^3} = \dfrac{T_M^2}{R_M^3} = k$ ?

Test:     does     $\dfrac{224.702^2}{108.21^3} = \dfrac{686.980^2}{227.94^3} = k$ ?

Calculate for Venus: $\dfrac{224.702^2}{108.21^3} = \dfrac{50490.98}{1267074.6} = 0.03984$

Calculate for Mars: $\dfrac{686.980^2}{227.94^3} = \dfrac{471941.5}{11842997.3} = 0.03984$

Hence k = 0.03984 in both cases! Try this calculation for other planets — even asteroids etc. (Exercise 22).

First we look at the ellipse from one point of view — one that a class at this level can grasp and model readily. This can serve as a first glimpse of the work on conics that we do in Year 9.

A good way to see this curve, the ellipse, is to compare it with a circle. It is easily constructed with a piece of string, and sticky tape, pencil and paper. There are many constructions for the ellipse (we do this in a *main lesson* in Class 9) but in this one, the dynamic quality of the form is apparent immediately as it requires *actual* movement — of a pencil — to make it happen.

*Exercise 20   Drawing the circle — no compass*

1. Obtain a piece of card about A4 size (30 cm × 21 cm), a pencil or coloured pencil with a long lead exposed, some fine string, and some masking tape or sellotape.

2. Place the card in landscape format

*Fig 4.39*

3. Draw a horizontal line across the centre and mark the middle.

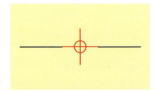

*Fig 4.40*

4. Now mark 1 cm distances on either side of the centre point *O* to 5 cm both left and right.

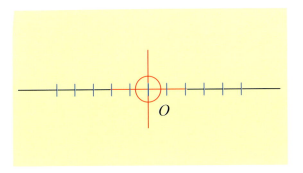

*Fig 4.41*

5. Take a piece of string about 15 cm long. Mark a distance of 2.5 cm from one end. From this mark, mark another point a further 10 cm along.

6. Pierce a small hole at *O*. Fold the string in half and feed through the hole from the back. Feed through until there is 2 times 5 cm in front. Tape the 2.5 cm ends at the back. This should leave a loop of string exposed on the front of 10 cm length.

7. Now the fun bit. With a sharp pencil so that the lead is just through the loop, tense the string so that the pencil point, *A*, is 5 cm from *O*. Keeping the string taut and the pencil vertical draw a path on the card. Draw this all the way round. What have we drawn? It should be a *circle* of about 5 cm radius.

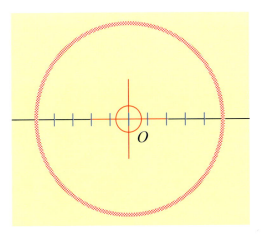

*Fig 4.42*

*Exercise 21   Modelling an ellipse family*

Now it gets interesting. What happens if we remove the tape
and pull out the string and re-insert the string (from the back)
through two *new* holes $F_1$ and $F_2$ each 1 cm away from *O?*

*Fig 4.43*

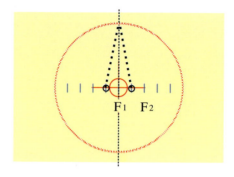

1. Make two holes at $F_1$ and $F_2$ respectively. Reinsert the string but
   this time through *both* holes so that a length of 10 cm is in front.

*Fig 4.44*

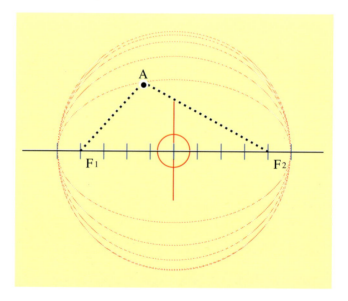

2. Again with a sharp pencil so that the lead is placed through the
   loop, tense the string so that the pencil point, *A,* is 5 cm from $F_1$
   and $F_2$ respectively. Again keeping the string taut and the pencil
   vertical draw a path on the card. Draw this all the way round.
   What have we drawn? It should be an *ellipse.*

3. What will be the minor axis and the major axis? Remember the string length *has not changed.* The vertical height (half the minor axis, or *semi*-minor axis), *b*, from *O* we can calculate using Pythagoras Theorem. It will be:

$b^2 + 1^2 = 5^2$.  What is *b*?

$b^2 = 5^2 - 1^2$, $b^2 = 25 - 1$, $b^2 = 24$, $b = \sqrt{24}$, $b = 4.8989$
$b = 4.90$ to 2 decimal places.

4. The length of *a* (half the major axis, or the *semi*-major axis) is 5 cm. Check this by simple calculation.

5. Now draw ellipses where $F_1$ and $F_2$ are both 2, 3, 4 and 5 cm from *O*. You should get figures like those shown in Fig 4.44, a family of ellipses all with the same major axis. (It would be a useful exercise to calculate the new *b* values for each of these new ellipses)

For those that need to know and perhaps to help with planning some work for students: If *c* is the distance between $F_1$ and $F_2$ (the inter-focal distance) then we find that $e = c/2a$ where *e* is the eccentricity of the ellipse, and *a* is the *semi-major* axis (usually shown horizontal) of the ellipse.

And if *b* is the semi-minor axis of the ellipse then $e^2 = 1 - (b^2/a^2)$.

For the example above we note that $c = 8$ cm and $a = 5$, so what are *e* and *b*?

Therefore if as given: $e = c/2a$ then $e = 8/(2 \times 5) = 8/10 = 0.8$

And also $e^2 = 1 - (b^2/a^2)$, then $0.8^2 = 1 - (b^2/5^2)$,

that is $0.64 = 1 - (b^2/25)$, or $b^2/25 = 1 - 0.64$

or $b^2/25 = 0.36$, $b^2 = 0.36 \times 25$, or $b^2 = 9$, so $b = 3$

Check the drawing to see that $a = 5$ cm (given) and $b = 3$ cm (as calculated).

This is enough theory about ellipses as such and gives enough control over the form for the moment. Kepler's second law is harder to deal with so we leave it as said and realize only that the nearer a planet (or comet) is to the Sun the faster it is travelling.

*Exercise 22   Kepler's third law*

From the table below:

| | MOON | MERCURY | VENUS | EARTH | MARS | JUPITER | SATURN |
|---|---|---|---|---|---|---|---|
| AVERAGE DISTANCE FROM SUN (in millions of km) | | 59.91 | 108.21 | 149.60 | 227.94 | 778.34 | 1427.01 |
| SIDEREAL PERIOD orbital time in DAYS | | 87.969 | 224.70 | 365.26 | 686.98 | 4332.59 | 10,757.20 |

*Fig 4.45  Distances of the planets from the Sun and average orbital periods*

1. Find the number of Earth days for a single orbit of the two inner planets, Venus and Mercury.

   For Venus            224.7, For Mercury        87.969,

2. Find the number of Earth days for a single orbit of the three outer planets, Mars, giant Jupiter and ringed Saturn.

   For Mars         687.98, For Jupiter        4 332.59,

   For Saturn        10 759.2,

3. Determine the *ratios* of these times in relation to Earth years, assuming 1 Earth year = 365.26 days (e.g. for Mercury it will be 87.969 ÷ 365.26 = 0.241 to three decimal places)

   Mercury : Venus : Earth : Mars : Jupiter : Saturn

   0.241    :        :    1  :        :        :

          0.614              1.88     11.86     29.456

4. Explore the third law. We are told by Kepler that $T^2 = kR^3$ where $T$ is the orbital period and $R$ is the mean or average distance of the planet from the sun, and $k$ is a constant determined by the units used. Use $k = 0.03984$.

*Example:* Check that for Venus: $T^2 = kR^3$ very closely given that $T$ = 224.701 days and that mean distance, $R$ = 108.21 millions of kilometers when k = 0.03984.

> Left-hand side (L.H.S) $T^2$ = 224.701 × 224.701 = 50 490
> Right-hand side (R.H.S)
> $kR^3$ = 0.03984 × 108.21 × 108.21 × 108.21 = 50 480
> so L.H.S = R.H.S very closely.

For MARS:

L.H.S $T^2$ =
686.98  × 686.98 = 471 941

R.H.S $kR^3$ =
0.03984  × 227.94 × 227.94 × 227.94 = 471 825

For SATURN:

L.H.S $T^2$ =
10759.2  × 10759.2 = 115 760 384.6

R.H.S $kR^3$ =
0.03984 × 1 427.01 × 1 427.01 × 1 427.01= 115 771 158.5

5. Suppose an asteroid, Ceres, has a mean distance of 375 millions of kilometers from the Sun. What would be its orbital period assuming it obeys Kepler's third law?

> $kR^3$  = 0.03984 × 375  × 375  × 375 = 2100937 = $T^2$.
> So   $T = \sqrt{2100937}$ = 1449 days.

## Connecting the big and the little

Not only are there big rhythms and little rhythms. There is some relation *between* the rhythms of the large and the rhythms of the small, the macrocosm and the microcosm — well, is there? Astonishingly there is in one, at least, *very* important case. It involves all of us, *and* the entire solar system.

## Human and cosmic rhythms

Relation of Platonic year to a human life to the number of breaths a day. This correspondence was pointed out by Rudolf Steiner — this was the first time I came across it. It is a linkage worth examining.

*Exercise 23   A correspondence of micro and macro rhythms*

1. With a watch or stopwatch each student in the class counts the number of times they breathe per minute.
      Breathes per minute  $B =$                          16 to 20?

2. Find average for the class of $n$ students.
      $x =$ (Sum of $B$ for class) $/ n =$

   This should be about 18, or a little less, for this age group.

3. Now calculate how many breaths this is per day.

   18 per minute(say) $\times$ 60 per hour $\times$ 24 per day $= 25\,920$.

   We note that this is about 26 000.

4. Now do a bit of research. There is a cycle that the Sun and background stars have. It is called the Platonic Year. How long is this according to the assessments of the astronomers?

   You could find a number of figures but they should be about 25 000 to 26 000

5. If the average human life were about three-score years and ten (3 $\times$ 20 + 10 = 70), or 70 years, although many are living well past this today, it could be, on the whole, that this is the length of a life of a human being on Earth. Suppose also a year is about 360 days. What then is 360 $\times$ 70?        25 200
      Let a lifetime be 72 years then 360 $\times$ 72?        25 920

*Fig 4.46*

platonic cosmic year

When the sun rises on the first day of Spring (ie. the vernal equinox) it is in a particular constellation. Each year, the place of sunrise at the vernal equinox moves a little bit along the zodiac. This means that in the course of time there is a gradual shift through all the zodiac constellations of the starry world. After a certain period of time, the place of Spring's beginning must again be in the same spot in the heavens and for the place of its rising the sun has travelled once around the entire zodiac. Astronomers have calculated that this journey of the sun takes approx. 25,920 years. This period of time is called the PLATONIC COSMIC YEAR and is also sometimes referred to as the PRECESSION of EQUINOXES.

Thus we find the same interval in the human being (microcosm) as in the largest interval, the macrocosm.

## 6. Make a table:

| Intervals | Calculations | Result |
|---|---|---|
| Breathes per day | $18 \times 60 \times 24 = ?$ | 25920 |
| Days per Lifetime | $360 \times 72 = ?$ | 25920 |
| Earth Years per Platonic Cosmic Year | | 25920 |

Not only do we have very large and very small rhythms but we now find that they are connected through the magic of number. And there is even an intermediate cycle in which we all partici-pate — our very life itself! Put these side by side and it all gets interesting — the human is surely the microcosm in the macro-cosm. Man is made in the image of God.

# Acknowledgments

I wish to acknowledge the many colleagues at Glenaeon Rudolf Steiner School who have conversed with me, also helped and challenged me with their expertise and interest. In our Mathematics faculty area this includes Chris Collins, Lynn Cooper and Matthew Wright. Among these is an early guide, Cedric Leathbridge. This acknowledgement also includes students, some named but some not known or not sure, whose work I have included. Also I want to include a number of folk whose conversation in these areas I always enjoy in particular our old 'Morphology' group with Christal Post, David Bowden and Roger McHugh (who has now left us). These have always supported these studies. Similarly the individual I always regard as my teacher, Lawrence Edwards (who died in 2003), to whom I am hugely in debt and owe the most, as well as to Nick Thomas and Graham Calderwood while Andrew Hill has encouraged. Anne Jacobsen has let me use a number of her sketches from her delightful main lesson work and I am grateful for this.

Past students Luke Fischer, and Rosie Goodman, Paul Beasly and Daniel Beasly, Yasmin Funk, Madelaine Dickie, Georgia van Toorn and Jenny Ellis have challenged and inspired. Not forgetting Marco, Annika and Claire. And Terry who got me some snails. And Ashley Miskelly whose sea urchin expertise astounds. And Elaine for her tree photos and introducing me to the Queen of the Night. And Ian Williams for weird meteorites. And very recently Anne Williams for woven hyperbolic figures which *must* be included in the next edition under the heading 'Holes and Crinkles.' These relevant models can readily be made by students who can even revive their crochet skills.

All these people have helped my research into these topics, into the patterning of things which — I believe — maths sort of *is*. I am also very grateful for the feedback from the last text which — so far — has been encouragingly positive!

Finally, I must thank my wife Norma for all the little things of nature which she has noticed, brought to my attention, and added to my store of perceptions and thoughts to work with. This is apart from her patience and probing questions. May these long continue!


John Blackwood

# Bibliography

Abbott, Edwin A. (1999, originally published 1884) *Flatland —*
   *A Romance in Many Dimensions,* Shambala, Boston and London
Alder, Ken (2002) *The Measure of all Things,* Little, Brown, London

Ball, Philip (1999) *The Self-Made Tapestry,* Oxford University Press
Bentley W A and Humfreys W J (1962 first published 1931) *Snow*
   *Crystals,* Dover Books
Blatner, David (1997) *The Joy of π,* Penguin, London
Bockemühl, Jochen (1992) *Awakening to Landscape,* Natural Science
   Section, The Goetheanum, Dornach, Switzerland
Bortoft, Henri (1986) *Goethe's Scientific Consciousness,* Institute for
   Cultural Research
— (1996) *The Wholeness of Nature,* Lindisfarne, New York, and
   Floris Books, Edinburgh

Casti, John L (2000) *Five More Golden Rules,* John Wiley, New York
Clegg, Brian (2003) *The First Scientist,* Constable, London
Colman, Samuel (1971, first published 1912) *Nature's Harmonic*
   *Unity,* Benjamin Blom, New York
Cook, Theodore Andreas (1979, first published 1914) *The Curves of*
   *Life,* Dover Books
Critchlow, Keith (1976) *Islamic Patterns,* Thames and Hudson, London
— (1979) *Order in Space,* Thames and Hudson, London
— (1979) *Time Stands Still,* Gordon Fraser, London, and St Martin,
   New York

Daintith, John and Nelson R. D., (1989) *Dictionary of Mathematics,*
   Penguin, London
Davidson, Norman (1985) *Astronomy and the Imagination,*
   Routledge, London
— (1993) *Sky Phenomena,* Lindisfarne, New York, and Floris Books,
   Edinburgh
Doczi, Gyorgy (1981) *The Power of Limits,* Shambala, Colorado

Eisenberg, Jerome M (1981) *Seashells of the World,* McGraw-Hill,
   New York
Edwards, Lawrence (1982) *The Field of Form,* Floris Books, Edinburgh
— (2002) *Projective Geometry,* Floris Books, Edinburgh
— (1993) *The Vortex of Life,* Floris Books, Edinburgh

Endres, Klaus-Peter and Schad, Wolfgang (1997) *Moon Rhythms in Nature,* Floris Books, Edinburgh

Folley, Tom and Zaczek, Iain (1998) *The Book of the Sun,* New Burlington, London

Gaarder, Jostein (1995) *Sophie's World,* Phoenix House, London
Garland, Trudi Hammel, *Fascinating Fibonaccis,* Dale Seymour, New York
Ghyka, Matila (1977) *The Geometry of Art and Life,* Dover Books, New York
Gleick, James (1987) *Chaos,* Penguin Books, New York
Golubitsky, Martin and Stewart, Ian (1992) *Fearful Symmetry,* Blackwell, Oxford
Goodwin, Brian (1994) *How the Leopard Changed Its Spots,* Weidenfeld and Nicholson, London
Guedj, Denis (1996) *Numbers: The Universal Language,* Thames and Hudson, London
Gullberg, Jan (1997) *Mathematics, From The Birth Of Numbers,* Norton, New York

Hawking, Stephen (2001) *The Universe in a Nutshell,* Bantam, London
Heath, Thomas L. (1926) *The Thirteen Books of Euclid,* Cambridge University Press
Hoffman, Paul (1998) *The Man Who Loved Only Numbers,* Fourth Estate, London
Holdrege, Craig (2002) *The Dynamic Heart and Circulation,* AWSNA, Fair Oaks
Hoyle, Fred (1962) *Astronomy,* Macdonald, London
Huntley, H. E. (1970) *The Divine Proportion,* Dover Books

Kollar, L. Peter (1983) *Form,* privately published, Sydney
Kuiter, Rudie H (1996) *Guide to Sea Fishes of Australia,* New Holland, Sydney

Livio, Mario (2002) *The Golden Ratio,* Review, London
Lovelock, James (1988) *The Ages of Gaia,* Oxford University Press

Maor, Eli (1994) *The Story of a Number,* Princeton University Press
Mandelbrot, Benoit B (1977) *The Fractal Geometry of Nature,* W. H. Freeman, New York
Mankiewicz, Richard (2000) *The Story of Mathematics,* Cassell, London
Marti, Ernst (1984) *The Four Ethers,* Schaumberg Publications, Roselle, Illinois
Miskelly. Ashley (2002) *Sea Urchins of Australia and the Indo-*

*Pacific,* Capricornia Publications, Sydney

Moore, Patrick and Nicholson Iain (1985) *The Universe,* Collins, London

Nahin, Paul J (1998) *The Story of √–1,* Princetown University Press

Pakenham, Thomas (1996) *Remarkable Trees of the World,* Weidenfeld & Nicolson, London

Peterson, Ivars (1990) *Islands of Truth,* W. H. Freeman, New York

Peterson, Ivars (1988) *The Mathematical Tourist,* W. H. Freeman, New York

Pettigrew, J. Bell (1908) *Design in Nature,* Longs, Greens and Co., London

Plato, *Timaeus*

Posamentier, Alfred S, and Lehmann, Ingmar (2004) *A Biography of the World's Most Mysterious Number,* Prometheus Books, New York

Richter, Gottfried (1982) *Art and Human Consciousness,* Anthroposophic Press, New York, and Floris Books, Edinburgh

Ruskin, John (1971, originally 1857) *The Elements of Drawing,* Dover Books

Saward, Jeff (2003) *Labyrinths & Mazes,* Gaia Books, Stroud

Schwenk, Theodor (1965) *Sensitive Chaos,* Rudolf Steiner Press, London

Sheldrake, Rupert (1985) *A New Science of Life,* Anthony Blond, London

Sobel, Dava (2005) *The Planets,* Fourth Estate, London

Steiner, Rudolf (1984, originally 1923) *The Cycle of the Year,* Anthroposophical Press, New York

— (1972, originally 1920) *Man: Hieroglyph of the Universe,* Rudolf Steiner Press, London

— (1960, originally 1922) *Human Questions and Cosmic Answers,* Anthroposophical Publishing Company, London

— (1991, originally 1914) *Human and Cosmic Thought,* Rudolf Steiner Press, London

— (1947, lectures given in December 1918) *How can Mankind find the Christ again,* Anthroposophic Press, New York

— (1961) *Mission of the Archangel Michael, 6* lectures given in Dornach, Switzerland, in 1919, Anthroposophic Press, New York, USA

— (1997, originally 1910) *An Outline of Esoteric Science,* Anthroposophic Press, New York

— *The Relation of the Diverse branches of Natural Science to Astronomy,* 18 lectures given in Stuttgart, Germany, in 1921

— (1983) *The Search for the New Isis, Divine Sophia,* Mercury Press, New York

Stevens, Peter S. (1974) *Patterns in Nature,* Penguin, New York

Stewart, Ian (1989) *Does God Play Dice,* Penguin

— (1998) *Life's Other Secret,* Penguin

Stewart, Ian (2001) *What Shape is a Snowflake?* Weidenfeld and Nicolson, London

Stockmeyer, E.A.K (1969) *Rudolf Steiner's Curriculum for Waldorf Schools,* Steiner Schools Fellowship

Strauss, Michaela (1978) *Understanding Children's Drawings,* Rudolf Steiner Press, London


Tacey, David (2003) *The Spirituality Revolution,* Harper Collins, Sydney

Thomas, Nick (1999) *Science between Space and Counterspace,* Temple Lodge Books, London

Thompson, D'Arcy Wentworth (1992, originally 1916) *On Growth and Form,* Dover Books


Van Romunde, Dick (2001) *About Formative Forces in the Plant World,* Jannebeth Roell, New York


Wachsmuth, Guenther (1927) *The Etheric Formative Forces in Cosmos, Earth and Man*, New York

Wells, David (1986) *The Penguin Book of Curious and Interesting Numbers,* London

Wolfram, Stephen (2002) *A New Kind of Science,* Wolfram Media

Whicher, Olive (1952) *The Plant Between Sun and Earth,* Rudolf Steiner Press, London

— (1971) *Projective Geometry,* Rudolf Steiner Press, London

— (1989) *Sunspace,* Rudolf Steiner Press, London


Zajonc, Arthur, (1993) *Catching the Light,* Bantam, New York

# Index